D1583623

ROYAL BOTANIC GARDENS

Kingdom of Plants

A Journey Through Their Evolution

Will Benson

FOREWORD BY Professor Stephen Hopper

DIRECTOR, ROYAL BOTANIC GARDENS, KEW

Collins

Collins
A division of HarperCollins*Publishers*
77–85 Fulham Palace Road
London W6 8JB

www.harpercollins.co.uk

Collins is a registered trademark of HarperCollins*Publishers* Ltd.

First published in 2012

Text © Atlantic Productions (Chevalier) Ltd 2012
Foreword © Board of Trustees of the Royal Botanic Gardens, Kew 2012
Photographs © individual copyright holders

17 16 15 14 13 12
10 9 8 7 6 5 4 3 2 1

A catalogue record for this book is available from the British Library.

ISBN 978-0-00-746333-6

Colour reproduction by FMG
Printed and bound by Gráficas, Spain

Contents

Foreword by

Professor Stephen D. Hopper

Plant diversity underpins most life on Earth, including ours. Plants provide the oxygen for every breath we take. Think of our daily ingestion of water and food, staying healthy through medicines, living comfortably in buildings, the joys of music, reading, gardening, farming, growing trees and exploring nature.

This book reveals some of the wonders of the plant world, exploring scientific information on the evolution of plants, the wonder of flowering plants, their diverse form and function, and their vital place on Earth. Stories abound of plants having a profound impact on our planet throughout evolutionary history, right up to the present day of unprecedented global change.

It is thrilling for me to revisit this great journey in these pages, from the remarkable evolution of photosynthesis in primeval oceans to the great move onto land through the evolution of desiccation-tolerant plant bodies, pollen and seeds. Who could not marvel at the diversity of the world's estimated 400,000 flowering plants? Few are not moved by the sheer beauty and intellectual challenge of understanding how such rampant life came to be. This book eloquently tells that story in an accessible up-to-date text.

For those whose interest is more excited by practical uses or the conduct of human affairs, there is much food for thought here as well. Apart from providing us with oxygen and moderating our climate, we learn that plants form a rich foundation for the web of life, and touch the lives of all people in every nation.

A special appeal of this book is its celebration of plant scientists and the insights they continue to bring. Charles Darwin devoted the best part of his last 20 years to the experimental study of plants, realising their value in illustrating evolution by natural selection, and publishing more books and journal pages on plants than on his more celebrated works in geology and zoology. Today, there are more plant scientists alive than ever before, and their discoveries in the field and laboratory are just as compelling.

Much remains to be done in the exploration of plant diversity. We are still discovering and naming 2000 new species of plant on Earth each year, from rainforest trees to colourful shrubs and orchids of temperate climates (especially in the southern hemisphere). Plant diversity unquestionably underpins human existence and livelihoods, yet we continue to destroy it at an alarming rate, with one in five plant species recently estimated to be under threat of extinction.

In our rapidly changing world, we are at a turning point for plant diversity. Without a fundamental shift towards more active conservation and the sustainable use of plants, our prospects for the future are grim indeed. Communicating this message in exciting and innovative ways to mass audiences remains essential. I am delighted that the Royal Botanic Gardens, Kew, has collaborated recently with Sir David Attenborough and Atlantic Productions in providing the plants and location for the astonishing series *Kingdom of Plants 3D*. I commend this programme and this beautifully produced book to all those who share a sense of wonder in plants and an appreciation of their central place in the lives of all on our one breathing planet.

Professor Stephen D. Hopper
Director (CEO & Chief Scientist)
Royal Botanic Gardens, Kew
April 2012

Kingdom of Plants 3D
with David Attenborough

'Watching flowers develop and insects visiting flowers is always wonderful to see, but in 3D it becomes transcendental. Seeing this happen in such depth gives you a most extraordinary, vivid impression. It's unforgettable.'

- David Attenborough

To preview the *Kingdom of Plants* app, scan the code above, and embark on your own journey with David Attenborough as he uncovers the hidden realm of plants at Kew.

This book, which explores both the splendour and diversity of the world of plants, was written to accompany the landmark television series *Kingdom of Plants 3D with David Attenborough*, produced by Atlantic Productions for Sky3D. Using cutting-edge stereoscopic cinematography, this revolutionary 3D project follows David Attenborough in his 60th year of broadcasting, as he uncovers the most amazing spectacles of the botanical world. Filmed over one year at the Royal Botanic Gardens, Kew, this series captures time-lapse footage of flowers as they burst into bloom, captivating interactions between plants and insects in microscopic detail, as well as snapping carnivorous plants and mystifying orchids – all filmed for the first time in three dimensions. Discover how plants first began to live on land, and how they have come to fill their place in the natural world today.

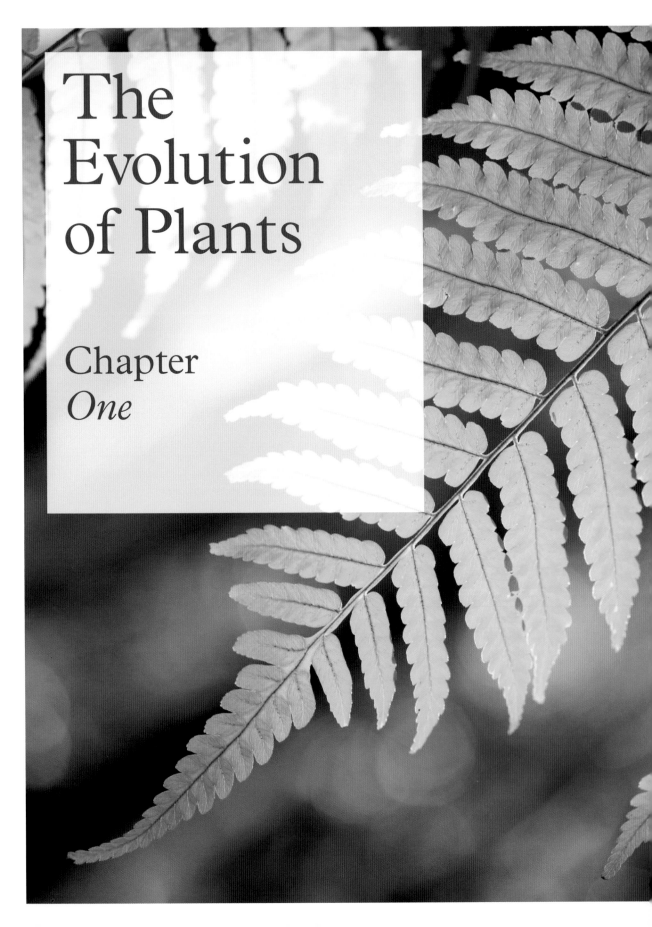

The Evolution of Plants

Chapter
One

'With the spread of plants onto land, their shapes and sizes soon began to diversify.'

In the last five hundred million years, plants have undertaken an epic evolutionary journey that has altered the very make-up of the planet. This journey began in a dark and acidic world and continues today in a world of rich colours, ornate shapes and mesmerising smells.

Every single step of their incredible evolutionary journey has been integral to creating the world we live in today. It is only by pulling apart the threads which create this rich network of flora that we can begin to uncover the extraordinary ways in which plants have come to live and thrive in all environments.

From their humble beginnings plants have become progressively complex and increasingly important to life on Earth. It is only through the success of the plant life on our planet that the animals that walk the Earth can be supported and it is the success of plants that has allowed humans to exist. Ultimately it is plants that will secure a place for our species to exist long into the future.

above: Blue morpho
(*Morpho peleides*)

Plants today help support vast
ecosystems of interconnected
species.

opposite: *Magnolia* sp.

One of the many species that were
brought to Europe and are now
common.

Over millions of years plants have become increasingly defined and specialised, carving out their own niches on the surface of the planet; each one striving for the evolutionary equivalent of the limelight – a chance to reproduce and spread their genes. In the beginning of the journey we uncover the origins of photosynthesis, an elegant mechanism which revolutionised the way organisms could obtain energy, providing a powerhouse through which plants could grow and compete. In the next step, plants make a crucial leap out of their watery beginnings, as they moved onto the land. But it is from here that the botanical world really came into its own. Plants became tough and tall, developing structural wood that allowed them to reach new heights. They developed mighty anchoring roots and broad, air-pumping canopies. Skipping forward a few hundred million years we meet the rise of the first flowers, signalling the beginning of a great and ground-shaking love affair between insects and plants, setting in motion a course of events that would bind some plants to the animal world. From this point, the pace of the story picks up as we enter the age of rich biodiversity and intricate connections between Earth's flora and fauna; we discover the extraordinary relationships between plants and animals, and we see that from one form of simple plant, many millions of diverse species have now evolved. And yet the most crucial chapters in the story of plants are still being written

From the early 1800s onwards countless expeditions set off to chart new territory and to collect rare and beautiful botanical specimens with which to fill elaborate glasshouses. This was to be the golden age of plant hunting, and with it began humankind's fascination with the plant world. Today the spirit of these first pioneers remains in the world's premier botanical gardens and research institutes. The biologists and ecologists who populate these centres have provided us with a far greater understanding of our planet's biodiversity than ever before. Ultimately it is these botanists and conservationists who will ensure the survival of the world's plant species.

★　★　★

above: Stromatolites

These fossils in Australia's Shark Bay represent the earliest known life forms on Earth.

—

To discover the origins of the plant kingdom we have to visit the Earth three billion years ago. At this time a dark gas-filled sky looms ominously overhead, the air is thick and acrid from the smouldering funnels of volcanic activity, and the waters of warm, shallow tropical seas lap at the shores of recently formed magma islands. This is a time we know now as the Archaean eon. In the watery depths of these ancient oceans tiny single-celled organisms drifted through the murky sediments of the seas. These basic microscopic cells were early bacteria, and consisted of nothing more than a simple outer membrane containing just a few primitive proteins inside. Over time, these cells grouped together to form layers of slime across the ancient seabed.

These bacteria survived by absorbing near-infrared light from the sun's rays that penetrated the ancient atmosphere. This light was used to convert the carbon dioxide and hydrogen-based organic chemicals that they had ingested from the water into sulphates or sulphur, providing them with nutrients. Although this basic chemical conversion may seem simple and insignificant, it was in fact the origin of all plant life we see on our planet today. This chemical conversion is the mechanism which all species of plant and animal on Earth, either directly or indirectly, now use as their ultimate source of energy. This was photosynthesis – the use of the light energy to manufacture vital organic food.

The first photosynthetic bacteria had bundles of light-absorbing pigments enclosed within their cell walls. These pigments were called bacteriochlorophylls, the predecessors of chlorophyll. The ability for these early cells to use energy from the sun to create organic compounds and sugars which could then be used for growth and movement was a major step forward in evolutionary terms. No longer would these Archaean bacteria be confined to absorbing the mere chemical scraps of nutrients available in the sediment. With their gradual radiation throughout the Archaean seas, the bacteria developed and adapted, and over hundreds of millions of years they evolved significantly.

Then, around 2.7 billion years ago, there emerged a further advance in these organisms' energy-exploiting capabilities. Alongside the early bacteria, new cells appeared – the cyanobacteria. Now while the early bacteria were making use of the 'invisible' near-infrared light from the sun, the structure of the pigments in the cyanobacteria's light-absorbing machinery had evolved to absorb visible light as a means of breaking down chemical compounds to produce food. To help them absorb this visible light even more efficiently they developed a far more varied range of photosynthetic pigments, called phycobilins and carotenoids, as well as several forms of what we know today as chlorophyll.

With the change in the wavelengths of light that these new pigments were able to absorb came a change in the precise chemicals that they could digest. For over 300 million years since the rise of photosynthetic bacteria, the by-product of photosynthesis had been sulphurous gases, but in the case of cyanobacteria the by-product was a simple yet vital molecule – oxygen.

So successful were the oxygen-pumping cyanobacteria in the prehistoric world that great colonies of them, many billions of cells strong, are now found fossilised in the layers of sediment which were laid down during the Archaean and Proterozoic eons. This record of life, captured as it was over two billion years ago, marks a critical juncture in Earth's history. As the cyanobacteria went about their business absorbing carbon dioxide from the sea and churning out oxygen into the water, some of this oxygen began to make its way up into the atmos-

phere, where it accumulated in great clouds, many thousands of tonnes in weight. At the same time – and for reasons still not conclusively known – others gases such as hydrogen began to decrease in the atmosphere. Crucially, this reduction in atmospheric hydrogen allowed oxygen to start accumulating.

Although they are not technically plants themselves, this so-called Great Oxidation Event earns cyanobacteria their place on the plant wall of fame. Oxygen is the third most abundant element in the entire universe, but until the intervention of these simple bacterial cells, Earth's oxygen atoms were predominantly locked up in chemical relationships with other elements. Elemental oxygen is so chemically reactive that whenever it gets the chance it will bond to nearly any other available molecule, and in the Archaean and Proterozoic eons, there was certainly no shortage of available hydrogen, sulphur and carbon which it could bind to, creating water (H_2O), sulphur dioxide (SO_2) and carbon dioxide (CO_2). But, critically, what cyanobacterial photosynthesis did was to split single oxygen atoms away from water molecules by using energy from sunlight, and then join two lone oxygen atoms together. By forming a partnership with an identical atom, these pairs of oxygen (or O_2, as you'll know them) had alleviated their want to bond to other chemicals through partnering up with one of their own. For the first time oxygen existed in a stable state, and this meant it could begin to increase in the atmosphere.

From the point, two billion years ago, at which cyanobacteria first began to produce significant amounts of oxygen by photosynthesis, it still took a further billion years for the levels of oxygen to reach even half of what they are today. Around 1.6 billion years ago, during the mid-Proterozoic eon, oxygen had risen to comprise about 10 per cent of the Earth's atmosphere, and by now the cyanobacteria had been joined by a host of varied photosynthesising life. These were the red algae, brown algae and green algae, and for the next 500 million years the soft and slimy, filamentous bodies of these organisms thrived in the oceans of the prehistoric world. As these algae evolved, they increased in complexity, developing advantageous new adaptations. Some developed multiple and specialised cells, allowing them to absorb nutrients and sunlight more effectively, and by dividing labour to different cells their growth and reproduction became more streamlined. Coupled with this, the various different algae packaged their genetic material into a single central nucleus, which distinctly separated them from earlier photosynthetic life. Unlike the cyanobacteria before them, these algae were among the first eukaryotic life forms, made of the type of complex cells that make up all higher life on Earth today. Most importantly, their light-absorbing chlorophyll pigments became stacked and enclosed within a double cell membrane, creating the self-contained photosynthetic structures we know from plants on Earth today. These are called plastids, or when in green algae and plants, chloroplasts. These relatively basic but increasingly ingenious organisms, although still confined to their aquatic habitat, first began to embody what we now recognise as the first plant life on Earth.

The vast colonies of red algae (whose colours actually ranged from green to red to purple to greenish black) stretched across the ocean floor, where they absorbed the shorter wavelengths of light that penetrated the murky depths. Brown algae soon became adapted to rocky coasts, attaching themselves to sub-merged rocks by structures called holdfasts. Crucially, green algae acquired an advantage that enabled them to thrive in the shallow waters of land formations. Unlike most other algae, green algae (known as Charophyceae) are able to sur-vive, and even flourish, in the strong light of exposed shallows. There are very few fossilised remains of the marine algae of this time, as their bodies were soft and easily broken down when they died. The exact stages that they underwent as they moved closer to land are unknown. However, we can deduce that around 500 million years ago green algae from the marine habitats and freshwater lakes washed ashore and became stranded on land. Some of these would have given rise to the first land plants. The earliest trace of plants on land that is recorded in the geological record is of the reproductive spores of a plant from the Ordovician period, some 470 million years ago. Analysis of these spores has revealed tiny structures which resemble those seen in a type of modern-day primitive plant called a liverwort.

The Palaeozoic era, which literally means the age of 'ancient life', stretched from 543 million years ago to 251 million years ago. From this time onwards scientists have been able to trace individual groups of early life. Around 500 million years ago, in a period within the Palaeozoic called the Cambrian, oxygen levels in the oceans dropped drastically, causing a condition called

anoxia, which soon spread across the planet. This may have been the trigger for algae to move from water to land. Many of the free-floating, bottom-dwelling organisms in the sea were killed, and this locked away thousands of tonnes of organic matter in their decaying bodies. As a result, the photosynthetic plankton increased in numbers to exploit the space and the nutrients, and they began to pull great quantities of carbon dioxide out of the atmosphere, in turn releasing large amounts of oxygen. Over a period of a few million years oxygen levels rose from around 10 to 18 per cent, up to as high as 28 per cent. This level has since fluctuated over subsequent geological history, resting today at about 21 per cent. So successful were the photosynthetic plankton that they still fill all corners of our oceans today. Just a single drop of water from the top 100 metres of the oceans will contain many thousands of these free-floating organisms. They are still considered to be some of the most important producers of organic matter on Earth.

above: *Marchantia polymorpha*

With no internal vascular system, liverworts rely on a moist external habitat.

overleaf: Cyanobacteria

These ancient microbes were the first to produce oxygen by photosynthesis.

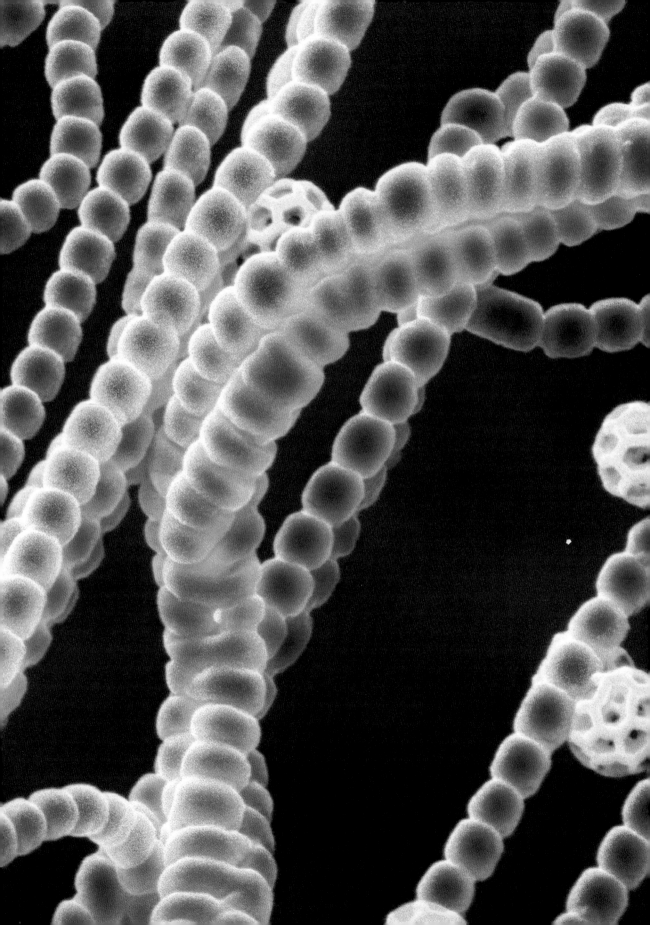

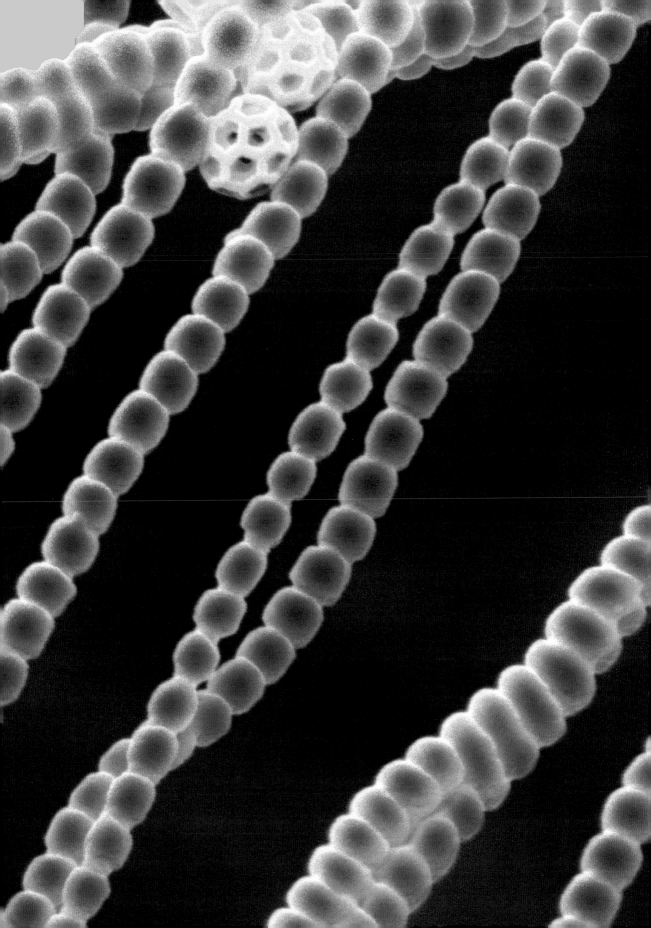

above: Bryophyte
spore capsules

Even on land these plants require
a partially wet environment for
reproduction.

—

Alongside the oxygen released from the oceans, the first land-based plants further increased the amount of oxygen that accumulated in the atmosphere. When the concentration eventually tipped over the crucial 13 per cent mark, the first wildfires became possible, and sparks caused by rock-slides and lightning set huge areas of the ancient landscape alight. Fossilised bands of these charcoaled plants, 430 million years old, have been found today. With an abundance of oxygen now readily available out of the water, and with competition for space and resources under the water increasing, life on land became a more favourable option. But while their soft, moist bodies were well suited to an aquatic life, the warm, dry air would cause their thin cell walls to quickly desiccate. More so, water was still necessary for their reproduction, in order to combine their male and female gametes.

Over a period of many hundreds of thousands of years, mutations occurred in the cells of some algae which gave them a chance to live further away from the safety of the aquatic environment. A waxy cuticle developed by some algae helped them resist desiccation, and gradually a layer of cells evolved to form a capsule around the embryo to protect it from exposure to the dry air. In time these desiccation-proofed algae reproduced, giving rise to plants better prepared to live out of water. While large blooms of green algae remained water-bound in lakes and oceans, those which had evolved to live for periods outside the water soon began to lose resemblance to their algal ancestors. The bryophytes, as they are now known, became the first land plants on the planet.

Even with their adaptations to terrestrial life, these small green hair-like bryophytes, which we now divide into the mosses, hornworts and liverworts, were still reliant on water, in the form of moisture from swamps and bogs, or dew. As they had only recently left their aquatic environment, in evolutionary terms at least, the bryophytes lacked the ability to carry water and nutrients from the soil to their upper parts, and therefore relied on their bodies to be covered in moisture. Once inside their cells the water had to then pass from cell to cell by the slow process of diffusion. As a result of this, even after 450 million years on Earth, bryophytes have remained small and inconspicuous, confined to the dark, damp habitats. Whereas the lives of their marine ancestors were largely commanded by the ocean currents, the first bryophytes developed primitive root-like structures, allowing them to be anchored to the soil. However, not only were the bryophytes key to all land plants we see today, they are now the third most diverse group of plants, numbering well over 10,000 species on Earth. These plants are closely coupled with many important biological and geological processes, including nutrient cycling in tropical rainforests, as well as playing a crucial role in insulating the arctic permafrost.

If the first major step in the story of plants was the development of photosynthesis, and the second was their establishment on land, then the third crucial stage was the development of their ability to grow from their limp origins, to become tall and tough, and to gain reproductive success over rivals. But from the origins of the early bryophytes some 450 million years ago, plants had to overcome two major obstacles, in order for them to diversify into the shapes and forms that we see today – how to get water and nutrients to all those parts that are not in contact with the soil, and how to support these parts without the buoyancy of water. The solution was found in the

above: Silurian landscape

The Silurian period, 444 to 416 million years ago, saw the evolution of the first vascular plants.

—

tracheophytes, or vascular plants as they are more commonly known.

The evolution of a vascular system was crucial to the plant world. Vascular plants have been the basis of all terrestrial ecology since their arrival on land. Among the first were a group of branching, 10-centimetre-tall plants called *Cooksonia*, which have been found in the layers of sediment that were laid down during the Silurian period, 444 to 416 million years ago, and are found most commonly today in the fossil fields of the Welsh Borders. *Cooksonia* had a simple structure, with no leaves or roots, but an internal system of tubes allowing them to move water from beneath the ground up to their photosynthetic structures, and to evenly distribute fuel throughout their branching arms. This hollow internal channel was created by open-ended cells along the length of their stems. In some cells photosynthetic ability was traded in order to take on a purely structural role in the plant. These cells lost their nucleus and their life-giving organs, and instead their walls became thickened with structural sugars, such as cellulose and a tough material called lignin. These vascular plants began to fortify their walls with woody lignin, which gives plants the structure and strength to sprout upwards, unsupported except by their own woody tissues, a key characteristic that separated them from aquatic plants. Over the next 350 million years, the vascular plants would eventually give rise to cycads, ginkgos, ferns, conifers, and ultimately all flowering plants.

By the time vascular plants began to make their mark on land, the story of the origin of plants had already spanned a vast timescale of over 2.5 billion years. Over this period the world had shifted during its fiery volcanic youth, deep in the Archaean eon, its skies became filled with life-giving gases during the Great Oxygenation of the Proterozoic eon, and it had played host to the first endeavours by plants to colonise the land in the Devonian period. By now the world was warm and humid and its surface was dominated by the ancient landmasses of Gondwana and Laurasia. The seas continued to support an increasing array of marine animal life, dominated by filter-feeding bryozoa and a diversity of prehistoric fish, and for the first time animals began to follow in the tracks of plants, and make their way out of the lakes and oceans and onto the land.

With the spread of plants onto land, their shapes and sizes soon began to diversify. By possessing tough lignin-enforced stems that allowed them to counter the force of gravity, vascular plants soon evolved into an array of new and fascinating forms. From the vegetation of the early Devonian, which consisted of small plants no more than a metre high, plants soon began to use their rigid stems to reach new heights. Fossils which have been dated between 407 and 397 million years old show evidence of plants which produced thickened body parts completely separate from their water-carrying internal tubes. These additional structures were the first examples of plants producing bark. As well as the emergence of woody body parts like bark, fossils from the Devonian reveal a whole host of novel structures that emerged at that time, giving this period its name of the 'Devonian Explosion'. Fossils from this time include plants such as *Archaeopteris*, which had frond-like leaves, and plants like *Drepanophycus*, which had metre-long roots that could reach nutrients deep in the soil. These first tree-like plants grew in vast numbers alongside rivers and

above: Ginkgo tree

These plants are living fossils, dating back to the Permian period, some 270 million years ago.

—

estuaries and began to give height to the first primitive forests, some growing up to 20 metres tall. Other woody plants from the Devonian include *Rhacophyton*, which is suggested to be the precursor of the ferns, an 8-metre tree with a large crown called *Eospermatopteris*, and a plant called *Moresnetia* which is thought to have been the forerunner to seed plants. The Devonian Explosion also gave rise to many familiar plant species. The 12,000 or so species of ferns that still thrive throughout our Earth's tropical and temperate zones today bear testament to their early success in the Devonian.

Animal life was also taking new shape in the warm Devonian climate. Along the damp forest floor millipedes scuttled through the organic mulch, and the first predatory animals, such as trigonotarbids, thought to be relatives of modern spiders, crawled through the undergrowth searching for a meal. In the seas great armoured fish equipped with powerful slicing jaws were quickly increasing in a variety of shapes and sizes, as they evolved to fill the expanding niches of the marine world at the time. In the Late Devonian, around 360 million years ago, the first of these fish made tracks onto land, giving rise to four-legged, air-breathing amphibians, such as *Hynerpeton*.

The Devonian flora had a fundamental impact on the very nature and substance of the land itself. As the tough, tall forest trees put down their networks of anchoring roots, they began to transform the hard and rocky substrate beneath them into hospitable and nutrient-rich soil. Prior to the plant development of the Devonian, the land surface of the Late Silurian was largely exposed bedrock, near-impenetrable to early root systems. The spread of the land plants of the Early Devonian aided the chemical weathering of rocks, helping break them down and release their mineral nutrients. The plants supplied organic acids from the fungi which colonised their roots, and together with acids given off by the decomposition of plant matter on the top of the substrate, these leached into the rock. This leaching softened the rock, enabling the roots to penetrate further into it, gradually breaking it up into smaller sediments. Over time, as organic matter from the surface was drawn deeper down into the ground, the soil depth progressively increased, allowing it to accommodate

longer roots below the ground, and in turn larger trees above
ground. As the soils of the Late Devonian and Early Carboniferous progressively deepened and became more developed,
plant growth was greatly enhanced.

above: *Archaeopteris*

Archaeopteris was one of the first
tree-like plants to appear in the
Late Devonian.

All of the early plants up until the Middle Devonian
possessed male and female gametes which required water,
in some form, for their fertilisation. As the original aquatic
plants had a totally submerged existence, their sperm cells

overleaf: A world of
colour

With the evolution of flowers, the
face of the Earth was transformed
forever.

could freely move through the water to fertilise their ovum cells. This method
of fertilisation put some obvious limitations on where they could survive, and
out of water their reproductive strategy would have been impossible. Although
bryophytes lived on land, they still relied on a partially wet environment to
transfer their sperm cells and spores to their female gametes. We know from

above: *Equisetopsida*

For over 100 million years, horsetails dominated the understorey of the Devonian, Carboniferous and Permian forests, growing up to 30 metres high.

—

bryophytes living today, such as mosses, that when their surroundings are saturated some species store up several times their own weight in water as a reserve, and they are also able to stop their metabolism if their habitat dries out for long periods. These water-dependent land plants were therefore best suited to colonise the damp tidal shores of lowland streams of the Devonian forests, and mosses and ferns can still be found to thrive in these environments today. The need of these amphibious plants to be linked to a moist external environment for their reproduction would have been very limiting in all but the dampest of habitats, and so any plant that was able to break this reliance on water would have had an immense advantage. In the drier terrain further from the shoreline there would have been an abundance of space, light and nutrients. Natural selection soon favoured plants with the ability to grow and reproduce in the dry air of these new habitats. Their trick to surviving in dry air was to package up their reproductive cells in desiccation-proof capsules that could carry them through the air. Capsules we know today as pollen.

The first pollen structures that evolved were tiny packages of genetic material, light enough to be carried on the wind to the female cells of a neighbouring plant. On reaching their destination they put out a little tunnel through which their sperm cells could swim down to achieve fertilisation. For the first time, male and female plant structures were able to swap their genetic information over large distances in the dry air. To maximise their dispersing ability many pollen-bearing plants grew taller, and in time the skies filled with airborne DNA from a multitude of pollen-spewing Devonian flora. Although plants would still require water for photosynthesis, it was now possible for them to colonise new, drier regions of the land. From the coastal forests, plants began to push further into the empty expanses of the ancient world.

As pollen plants began to spread their domain further inland, and it became necessary for their gametes to travel over even greater distances to achieve pollination, a further major shake-up occurred in the way in which plants reproduced. This was one of the most dramatic innovations in the evolution of plants on land – the evolution of the seed. The earliest plants which exhibited seed-like structures are known as the progymnosperms, dating back to around 385 million years ago. They included trees like *Protopteridium*, and the leafy, 10- metre-tall *Archaeopteris*. The fossils of the trees from this period indicate that some, but not all, possessed structures resembling primitive seeds, suggesting that this was a time when the future of the seed hung in the balance. Like all plants before them, progymnosperms produced spores, but uniquely they were able to produce two separate types – micro-spores and mega-spores. This trait, called heterospory, suggests that progymnosperms were the most likely antecedents of all seed plants. Their ability to create variable spores is thought to have been the crucial intermediate evolutionary stage between plants with free-floating single spores and those with true seeds containing a spore-borne embryo.

The first true seed plants, which descended from the progymnosperms over 350 million years ago, were a group of tree-like ferns called pteridosperms, belonging to the major division of plants called gymnosperms. The word gym-nosperm literally means 'naked seed', as they produce seeds which are not fully enclosed in an ovary. In earlier seed-less plants, the gametophytes were released outside the parent plant, but in the pteridosperms the gametophytes were microscopic in size and retained inside the reproductive parts of the plant. This created a moist ovule in which fertilisation could take place, in essence creating a plant within the parent plant. Coupled with this, these embryonic packages were encased with some starting-off food, meaning that they could be transported, ready to germinate as soon as they found themselves in the right conditions. The protective packaging of these seeds also enabled them to remain dormant after dispersal, and wait until conditions were perfect to grow. This prevented the precious genetic material contained within from being wasted in times of flooding or drought.

Today seed-bearing plants are the most diverse group of all vascular plants. The evolution of the seed enabled the proliferation of land plants on the wind, in the water, along the ground and in the stomachs of animals. During the Carboniferous and Permian periods, the gymnosperms evolved prolifically, with their extant relatives today including conifers such as pine, spruce and fir, with their needles; ginkgos, with their fleshy seeds; and cycads, with their large palm-like leaves and prominent cones.

Around 300 million years ago a global ice age hit the planet, and the Earth became progressively drier and cooler as great bodies of ice formed at the poles and locked away precious water vapour from the atmosphere.

The reduction of atmospheric moisture caused vast areas of tropical forests and swamps to shrink and dry out, and with their ability to disperse their seeds and colonise drier environments, gymnosperms soon replaced ferns as the dominant plants on the planet. In time the higher-altitude regions of the planet became regions of cold-climate peat lands and swamps, which would have resembled something similar to the boreal taiga of modern-day Siberia. In the milder lowlands, deciduous swamp forests were dominated by the seed ferns of *Glossopteris* and *Gangamopteris*, along with large clubmosses and immense horsetails.

By the end of the Permian period the main continents of Earth's land masses had all fused together into one supercontinent called Pangaea, and parts of the planet had become arid with little rainfall, creating extreme desert landscapes. As deserts expanded and coastlines shrank, this extreme climate shift began to push many life forms to the brink, and by 248 million years ago, 95 per cent of the plant and animal species that had evolved by this point were wiped out. This marked the largest extinction ever known, and for the next 500,000 years complex life on Earth teetered on the brink of complete extinction. The 5 per cent of life that remained was sheltered from the extreme climate, in habitats that remained temperate and moist enough. These pockets of life harboured the fundamental DNA that had evolved so far. Over the following 50 million years, as the global climate became more amenable once again, plant life would bounce back to colonise the planet. Slowly plants began again to create temperate woods, tropical forests and dry savannahs.

As the Jurassic swamps and prehistoric woodlands began to spring back to life, plants continued to increase and diversify. Seeds, leaves and pollen became more specialised, and the world of plant life provided an abundance of food for the dinosaurs. Plants gave rise to fast-growing bamboo and shade-giving palms until 140 million years ago, when the plant world would be changed completely.

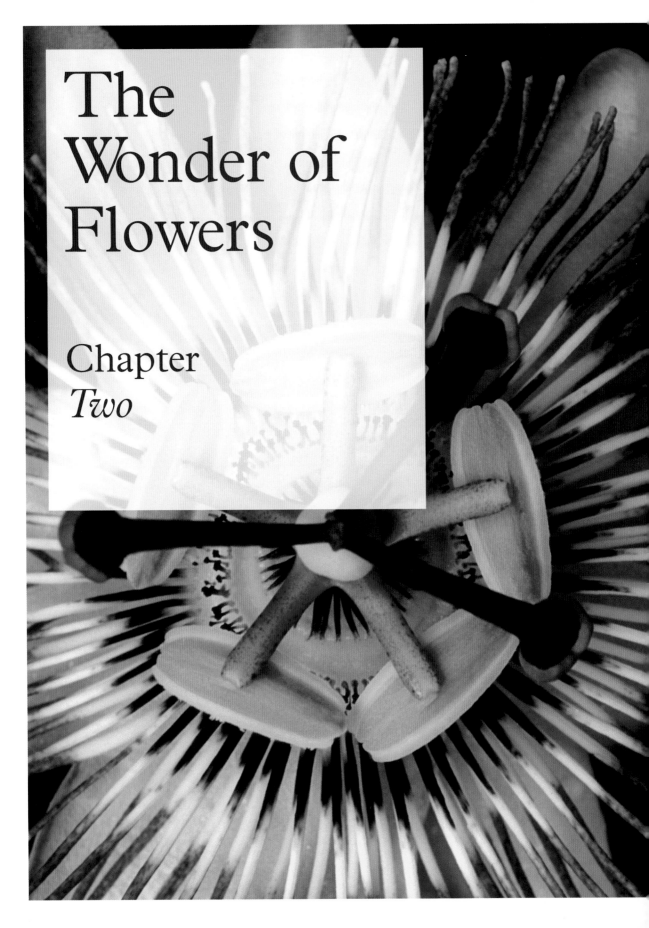

The Wonder of Flowers

Chapter
Two

'The botanical gardens and private collections of Europe's cities were soon overflowing with an explosion of fascinating and rare flowers.'

For more than two hundred years humans have had an obsession with flowers. It has seen men give their lives in search of the most exquisite floral specimens, and caused many others to lose their minds in pursuit of the rarest. The Victorians used the term orchidelirium to describe 'flower madness', the botanical equivalent of 'gold fever' for the 1800s. This fascination with exotic flowers began with the pioneering plant hunters of the eighteenth century, who sailed to South America, Asia and Africa, travelling through unmapped territory in search of botanical wonders.

These early expeditions were commissioned by wealthy collectors and botanical organisations, and they aimed to supply high society's increasing appetite for new and exciting plants and flowers. Often spending many years abroad at a time, plant hunters risked their lives, negotiating wild animals and hostile natives, in order to discover new plant species. The finest specimens could fetch a mighty price for their scientific uses and aesthetic value.

Our attraction to flowers has a deep history; evidence from a Neanderthal burial site in Iraq suggests that even 200,000 years ago our close hominid relatives were using flowers in ceremonies, laying the blooms from plants such as ragwort and grape hyacinth over the bodies of their dead. Throughout Greek myth flowers were sacred to both gods and mortals: the deep red of poppies was created from the drops of blood that fell from the slain Adonis, and the nymphs that sun-god Helios banished for their disloyalty were turned into the flowers of hellibores. In ancient Egypt roses in particular were a symbol of wealth, beauty and seduction. Guests at Emperor Nero's great banquets were showered with their petals, and it is documented that Cleopatra used the sweet scent of rose petals to lure Mark Antony. Flowers remain a huge part of our culture today, accompanying us on the most important days of our lives – our birth, our graduation, our marriage, our death. Our gardens are now awash with bright and showy blooms from habitats from all corners of the planet – magnolias from China, geraniums from the Cape of South Africa, primulas from the Himalayas and wisteria from the Orient.

In 1768 a botanist and horticulturalist named Joseph Banks set off with Captain James Cook on his first major voyage to the Pacific, where he would spend the next three years collecting, studying and cataloguing the wealth of fascinating new plant species that he found thriving on the tropical islands there. Following his return to England in 1771, Joseph Banks acted as an adviser to the Royal Botanic Gardens at Kew, a position that was later formalised. Banks gathered together a team of like-minded botanists and explorers for further expeditions. His team included the explorer and plant collector Allan Cunningham, and Scottish botanist Francis Masson, who would later become known as Kew's first plant hunter, and who would later join Captain Cook on his second major voyage. Under Banks' supervision the gardens at Kew fast became the world's foremost botanical garden. Impressive proteas, cycads and bird of paradise flowers from South Africa soon filled the greenhouses, with each species transported on its voyage enclosed in a mini-greenhouse, called a 'Wardian' case. It was the showy blooms and delicate scented flowers which drew the most attention back home in Britain. As the plant-collecting voyages pushed deeper through the thick vegetation of tropical jungles, increasing arrays of floral shapes and colours were collected, and made their way back to the collections at Kew.

Continuing Banks' legacy, his successor William Hooker, and later his son Joseph, maintained Kew's spirit of exploration, leading further trips to the mountains of India and Nepal. Among other species they discovered a mass of stunning new species of rhododendrons, a plant popular with gardeners across the world today. However, their plant-collecting exploits weren't always trouble-free, and during one of their trips to the Himalayas between 1847 and 1849, Joseph and his travelling companion Archibald Campbell were arrested and imprisoned for having illegally crossed the border from Sikkim

above: Sir Joseph Banks

Under his supervision, Kew's expanding collections of exotic plants saw it become a garden of international importance.

—

above: Sir Joseph
Hooker

Pen and ink portrait by
T. Blake Wirgman, 1886.
—

into Tibet. The two men and their botanical specimens were only released when the British government threatened to invade Sikkim.

As well as the public botanical collections of the time, such as Kew, obsessive private collectors also set out to acquire rare and exotic or even undiscovered flowers, which was lucrative for the financiers and explorers alike. The expeditions were often shrouded in secrecy to prevent rival groups from acquiring information as to where new species were likely to be growing, and it wasn't uncommon for false maps and information to be circulated in order to disorientate the competition. This was the age of orchidelirium, and successful collectors could sell their prized specimens at auction for colossal sums of money. It was these privately financed trips which brought back the first orchids to Britain, from the East, and in 1852 some of them made their way into the hands of London wine merchant John Day. Bought for the equivalent sum of £3000 in today's money, Day's first orchid flowers marked the beginning of a lifelong obsession. His house in north London was soon transformed with the delicate white and maroon petals of *Dendrobium* from South-east Asia, *Odontoglossum* from tropical America and *Cattleya* from Costa Rica. Combining his love of orchids with his keen artistic eye, Day set about documenting his increasing collection of flowers in a set of watercolours. His meticulous paintings, complete with notes on the plants' habitat, conditions for cultivating them and their price at auction, soon caught the eye of botanists and art lovers alike, and he was given special access to the orchid house at Kew to paint its plants. Over 25 years, Day compiled over 50 sketchbooks filled with his detailed, colourful visions of these captivating plants, and these drawings can still be admired in the collections at Kew today.

above: Plant hunting

From Joseph Hooker's *Himalayan Journals*, 1854.

—

The Victorian obsession with acquiring the most ornate flowers was made all the more possible by an extraordinary network of vivacious plant fanatics, who were willing to use their work in the far corners of the British Empire as an opportunity to bring back exotic species from across the globe. Colonel Robson Benson, an officer in the British forces in India, used his time on duty in Assam, Bhutan and Cambodia to collect a multitude of new species of orchid for the British horticulturist Hugh Low. Painter William Boxall, working first in Burma and later in the Philippines, collected enchanting slipper orchids, magnificent *Vanda*, and a number of species of the genus that today fills the shelves of nearly every garden centre, *Phalaenopsis*.

The botanical gardens and private collections of Europe's cities were soon overflowing with an explosion of fascinating and rare flowers, displaying an unfathomable array of shapes, sizes and colours. But as well as the aesthetic interest that drew most admirers to these flowers, their complexity and diversity provided biologists and naturalists with a wealth of material for them to study. One such naturalist was the young Charles Darwin, as well as Kew's second Director, Joseph Hooker, who was a lifelong friend of Darwin.

opposite: Illustrations by John Day, taken from his 'scrapbooks'.

(a) *Cattleya skinneri*

A species of orchid found in Costa Rica and Guatemala.

(b) *Catasetum christyanum*

An epiphytic orchid from northern South America.

(c) *Vanda coerulea*

A species of orchid discovered in Sikkim by Joseph Hooker in 1857.

(d) *Dendrobium formosum*

A species of orchid first discovered in northeast India.

Darwin shared extensive correspondence with a long list of senior botanists and horticulturists at Kew, swapping notes on plants and exchanging specimens. During his time on the *Beagle* between 1831 and 1836 he gathered species of flowers from Argentina, Chile, Brazil and the Galapagos which he sent back to Kew for identification, and in turn Kew happily provided Darwin with plants for him to document and study at his house in Kent. Although at this point Darwin had not yet written his seminal work *On the Origin of Species by Means of Natural Selection,* he was already piecing together his ideas on survival and adaptation in the natural world. Perhaps more than anyone else at the time, Darwin knew that for all their beauty, the complex shapes, patterns and structures of every unique orchid flower must be a result of some advantage that they bestowed upon that species in its habitat. Darwin understood that the flowers of orchids were purely about coaxing animals to spread its sex cells.

Darwin's instincts as a naturalist stemmed from his love of collecting, and his belief that in order to understand any aspect of the natural world, one must acquire, and carefully examine every facet of it. He once wrote, 'By the time I went to school my taste for natural history, and more especially for collecting, was well developed. The passion for collecting, which leads a man to be a systematic naturalist, a virtuoso or a miser, was very strong in me, and was clearly innate, as none of my brothers or sisters ever had this taste'. In his quest to make sense of the elaborate flowers of the orchid family Darwin began amassing his own collection of these rare plants, which he held in his glass conservatory at Down House. Countless orchids from Malaysia, the Philippines and Central America made their way via Kew to his house, together with the British species which grew in abundance nearby. But the nature of the most extreme orchid flowers did not fit well with his theory of evolution. In one letter that Darwin wrote in 1861 to John Lindley, who worked as one of Kew's taxonomists at the time, he describes his utter fascination with the complexity of orchids, discussing one genus in particular called *Catasetum*: 'I have been extremely much interested with *Catasetum*, and indeed with many

(a)

(b)

(c)

(d)

above: Deception

The labellum of this mirror bee-orchid has evolved to mimic the shape and shine of an iridescent bee.

—

exotic orchids, which I have been looking at in aid of an opusculus, on the fertilisation of British Orchids. I very much fear that in publishing I am doing a rash act; but Orchids have interested me more than almost anything in my life. Your work shows that you are carefully understanding this feeling.'

Darwin studied the lives of orchids and dissected them, looking at the multitude of ways in which the plants guided specific bees or moths to their flowers to interact with their reproductive structures, and the mechanisms they exhibited to achieve pollination. He was searching for an explanation for all aspects of each flower's behaviour, and a justifiable origin for each. But for many of his adversaries, what Darwin was trying to achieve was considered impossible, and even his good friend Thomas Huxley famously stated, 'who has ever dreamed of finding a utilitarian purpose in the forms and colours of flowers?' Darwin made good headway in unravelling the sex lives of orchids, and he made detailed studies of the ways in which they lured pollinators and released their pollen. But what had him most stumped was that, more so than any other family of flowers, orchids exhibit extreme pickiness in whom and what they allow to spread their pollen. On the subject of this he wrote: 'Why do orchids have so many perfect contrivances for their fertilisation? I am sure that many other plants offer analogous adaptations of high perfection; but it seems that they are really more numerous and perfect with the Orchideae than with most other plants.' What seemed counterintuitive to Darwin was that for all their elegance, the pollination methods employed by orchids seemed terribly inefficient.

Darwin's trouble with trying to explain the nature of the sex life of orchids becomes all the more apparent as soon as you begin to unfold the highly specialised ways which we now know different species achieve pollination. In the mirror bee-orchid (*Ophrys speculum*), found in southern and western Europe, as well as Lebanon, Turkey and North Africa, the lip of the flower looks nearly identical to an iridescent bee. This was first suggested to be a ploy to prevent grazing animals from munching on it, but we now know that it in fact releases a chemical that mimics the pheromones of a female bee, as a trick to get males to 'mate' with it. By rubbing its body on the flower in an attempt to copulate, the male bee will rub itself up against the plant's sticky bundles of pollen, called pollinia, which adhere to its body, before it flies away and attempts to mate with another bee-orchid. Another extreme behaviour has evolved in the orchids of the genus *Oncidium* from Ecuador, which have petals that look like the insect competitor of the *Centris* bee. The bee attempts to chase this 'enemy' away from its territory, and in doing so it strikes the flower, showering itself in the plant's sticky bundles of pollen. The slipper orchids of Asia and South America have a hinged lip which forces insects to brush past the sticky pollen before leaving, and there is even an underground species of orchid from Australia called *Rhizanthella slateri* which relies on ants to move its pollen. Other orchids emit a smell of rotting flesh to attract meat-loving flies to pollinate them, while some have been found to smell like chocolate.

Perhaps the most intriguing pollination syndrome is that of *Catasetum*, which is so complex it seemed to contradict Darwin's very theory of evolution. Unusually for orchids, some *Catasetum* plants are male and others are female. The male produces a scent that attracts just one species of euglossine bee. Lured by its sweet smell, the bee lands on the lip of the orchid and thrusts its head into the flower, touching a hair-trigger. This activates a mechanism that fires out a tiny bundle, which then sticks onto the bee's back. This extraordinary projectile is in fact a bundle of pollen grains called a pollinium, which has a little cap on it, and after a minute or so the cap falls off to reveal a little horseshoe-shaped bundle of pollen grains. A group of researchers in the USA recently found that the pollinium is ejected with an acceleration rate of over ten times that of a striking pit viper. Having been struck by this pollen, the bee flies away and is attracted to another rather different-looking flower, which is the female. Once again, lured by the scent, it sticks its head into the female flower – and the little bundle of pollen attached to its back, like a key, fits into a small aperture on the roof of the flower, like a lock, pulling off the pollen as the bee makes its departure. Pollination has been achieved.

Darwin's obsession with *Catasetum* in particular caused him to dedicate a great deal of time to studying the flower's mechanism, in an attempt to make sense of how it could have arrived at such a precise and specific system. His tireless persistence paid off, and in a letter to his publisher John Murray in 1861 he wrote, 'I have had the hardest day's work at *Catasetum* and the buds of *Mormodes*, and believe I understand at last, the mechanism of movements and functions. *Catasetum* is a beautiful case of slight modification of structure leading to new functions.'

Having unravelled the complexities of exploding pollen bundles, it wasn't long before Darwin's next botanical mystery would land on his desk, quite literally. In 1862 he received a package from renowned horticulturist James Bateman, a striking orchid with a flower composed of large star-shaped white petals from the island of Madagascar, named *Angraecum sesquipedale*. Darwin set about detailing the ornate nature of his latest specimen and was struck by the long tubes, called spurs, in which the plant's nectar was contained. Its delicate spurs were over 30 centimetres long, hanging down beneath the flower like white

opposite: The *Catasetum* conundrum

The structures of these orchids interested Darwin immensely on account of their incredible complexity.

—

opposite: Mystery
of the moth

*The relationship between hawk moth
and Angraecum sesquipedale is one
of the greatest examples of plant and
animal co-evolution.*

___

tails, with the nectar contained at their tip. Having never seen anything quite like this before, he wrote to his esteemed friend Joseph Hooker to explain these foot-long, whip-like nectaries, exclaiming, 'Good Heavens what insect can suck it!' Later that year Darwin went on to publish a book on the reproduction of orchids, in which he theorised that in order for the Madagascan orchid to be pollinated, an insect, most probably a moth, must exist on the island of Madagascar with a tongue at least 30 centimetres long which can reach the nectar at the end of the spurs. His suggestion seemed ludicrous to many of his peers, but a paper written by fellow evolutionary theorist Alfred Russel Wallace a few years later sought to back up Darwin's notion, by highlighting that a large hawk moth had been discovered in Africa which had a tongue almost 20 centimetres long, called *Xanthopan morgani*. Wallace predicted that if such a moth existed in Africa, then surely a moth with a 30-centimetre-tongue could live in the forests of Madagascar. Unfortunately Darwin was never able to see his prediction come true, but in 1903 a population of hawk moths with the necessary tongues were found on Madagascar. The team who discovered it then aptly named it *Xanthopan morganii praedicta* – the predicted subspecies of *X. morgani*.

But for all the sense that Darwin was able to make of the lives of flowering plants and how they disperse their pollen, there was one fundamental aspect of their world that he was never truly able to fathom. He couldn't understand how flowering plants had come to exist in such diversity in such a short period of geological history. For almost half a billion years, plants had existed without flowers, and then in a few million years they appear in the fossil record as the dominant form of plant life. In a letter sent to his friend Joseph Hooker in 1879, Darwin remarked on his puzzlement on the sudden radiation of flowering plants, stating, 'the rapid development, as far as we can judge, of all higher plants within recent geological time is an abominable mystery.' It went against his very theory of 'natural selection' that he had outlined for the rest of life on Earth.

Darwin's work, which observed the processes by which all species struggle for survival and compete to reproduce in order to pass on their genes, helped build his theory of evolution – the process by which a species over many generations acquires novel and advantageous traits. From his acute observations of the birds, insects and reptiles of the Galapagos Islands, together with the fossils of extinct animals that he gathered during his voyage on the *Beagle*, Darwin discovered that the change inferred to organisms over time was a slow and gradual process. In his subsequent book *On the Origin of Species by Means of Natural Selection* he states that 'natural selection acts only by taking advantage of slight successive variations; she can never take a great and sudden leap, but must advance by short and sure, though slow steps.'

However, for all the diversity that Darwin could see in the modern world of flowering plants, the fossil record revealed no trace of the expected slow and gradual transition of non-flowering plants to those with flowers. For Darwin, the seemingly sudden appearance of flowers contradicted the very rules of evolution. In the fossils of the Carboniferous period horsetails and early seed plants dominated the land. In the fossils from the time of the dinosaurs cycads, ginkgos and ferns were dominant. And then suddenly in the fossils of the Cretaceous period, 130 million years ago, an explosion of flowering life appears and takes over the land. In this short period all of the major groups of flowers that we see today emerged. Not only did each species of flower look different, but they had developed a multitude of reproductive styles – from the relatively straightforward mechanism of those flowers which released their pollen on the wind, to those possessing nectar-filled organs for the more complex task of luring insect pollinators. This sudden spurt of evolution not only had Darwin stumped but has continued to boggle botanists for the last 120 years.

Darwin knew that the fossil record was not a wholly complete snapshot of life through time, and he used this to try and explain the sudden burst of flowering plants on Earth. Due to the fact that plants do not have hardened body parts like the easily fossilised internal and external skeletons of many animals, there is a chance that the intermediate stages of the first flowers may simply have decomposed and been lost when they died. In his correspondence with Hooker, Darwin suggested that flowering plants had perhaps evolved slowly and that the fossils were yet to be found. Another suggestion was that a rapid increase in flower-frequenting insects in the Cambrian may have spurred their evolution on. In the animal world it is possible to see many of the intermediate steps that have given rise to certain animals. The embryos of snakes, dolphins and whales all sprout the buds of vestigial legs when they are embryos, echoing their evolutionary past, which then shrink and disappear before they are born. However, plants do not retain these evolutionary features in the same way that animals do, and it is far harder to trace the steps by which flowers came to be by studying their more primitive relatives. What makes things harder still is that the flowers of even moderately primitive groups of flowering plants are so different from their assumed extinct relatives among seed plants, that it is incredibly difficult to reconstruct a plausible evolutionary history for them.

However, since Darwin's day new fossil finds and our considerable advances in genetics have helped us begin to unravel the origins of flowering plants. In the mid-1980s an international collaboration of over 40 scientists from around the world, coordinated byKew geneticist Professor Mark Chase, embarked on a mammoth project. Over a number of years the team meticulously extracted the same type of gene from over 500 different types of flowering plant, and by the early 1990s they had gathered enough information to begin to compare them. By looking at the plants which shared the most similarities for this type of shared gene, Professor Chase and his team were able to work out which groups of flowers were more closely related, and by pinpointing those in which the gene had considerable differences they could assume they had evolved separately. The findings, published in 1993, allowed them to piece together an accurate tree of life for flowering plants. Fifteen years later a team of researchers at the University of Florida built on this tree of life to create a more complex timeline for the emergance of different types of flowers. By looking at the genes of a number of living plants that could be linked to their fossil ancestors from known dates in prehistory, the team were able to work out the rate at which certain genes changed over time. The results from these calculations gave them a ticking genetic clock which could then be used to date the origins of the first flowering plants. The major revelation from their work was that previous estimates for the first flowers had been inaccurate by around 10 million years, and that the first blooms were in fact evolving as far back as 140 million years ago. But the Florida team's findings still seemed to indicate that flowering plants did indeed rapidly radiate in as little as just five million years.

For years the debating and theorising continued as to how plants could have seemingly cheated evolution, to quickly rise from relative obscurity into a wealth of developed flowering structures during the Cretaceous period. Then scientists believed they had found an explanation, in the form of a happy coincidence, a genetic mishap discovered in some plants known as polyploidy. It has been known that when the male and female haploid sex cells of both plants and animals combine during reproduction to create the next diploid generation, some sections of genetic information from the parents can become duplicated in the new generation. In the case of humans, the accidental insertion of any additional genetic information can be extremely damaging for the child's health; even the duplication of just one of our 46 chromosomes will cause Down's syndrome, and the duplication of two or more chromosomes would be fatal. However, flowering plants with their comparatively simple body parts have been found to be able to live healthily with accidental duplications of genetic material, even in extreme cases where the whole of a plant's genome (i.e. every one of its chromosomes) becomes duplicated. Not only can a plant species tolerate these polyploidy events, but it appears that they can actually thrive on them.

It had long been considered that flowering plants' ability to duplicate large parts of their genetic information could have been a contributing factor that allowed them to increase in diversity at an abnormal rate, and a study carried out in the 1970s calculated that between 30 per cent and 80 per cent of all flowering plants have undergone a multiplication of parts or the whole of their genome at some point in their evolutionary history. Plants that have undergone polyploidy are typically more vigorous. By tracing through the lineage of many different groups of flowering plants, scientists have now found proof that a number of polyploidy events does indeed explain the fast rate at which particular types of flowers radiated, such as the prolific grasses, the nightshades, the pea family and the mustard flowers. But what of all the other flowering families that make up the 400,000 or so species on Earth today?

Up until very recently speculation still remained as to what exactly could explain the apparently sudden explosion of all flowering life 140 million years ago. Then in 2011, at a conference of the International Botanical Congress in Melbourne, a palaeontologist from the Swedish Museum of Natural History called Else Marie Friis revealed findings which outlined a previously unseen

trove of exquisitely preserved primitive flowers from the charcoal fields of Catefica,
Torres Vedras and Famalicao in Portugal,
dating back to the Early Cretaceous. These

above: *Nuphar lutea*

These yellow water-lilies are among
the oldest living relatives of
the first flowering plants.
—

amazing fossils, many of which were preserved in three dimensions, gave the
first glimpse of what early flowers looked like as they began to evolve, and
in breathtaking detail they showed the first stages of flowering life on Earth.
Some possessed clusters of small flowers grouped together to form one larger
inflorescence, much like a modern-day sunflower, while other plants had small
single flowers no more than 2 millimetres across. Most seemed to have few
floral parts, and many even lacked petals and the protective outer sepals which
are characteristic of most modern flowers. Numerous seeds and pollen were
also found in the fossils, and the high number of fruits possessing fleshy coats
suggests that animals had a key role in dispersing the seeds of these plants.
Friis's fossils seemed to reveal what flowering plants looked like some 30 million years before the fossil evidence of Darwin's day, at a point when they
were first acquiring the features which would ultimately lead to the flowers

above: Buzz-
pollination

The stamens of *Gustavia longifolia*
only release their pollen when
buzzed by the wings of a bee.

—

we see today. While historic polyploidy events undoubtedly gave ancient flowering plants occasional moments of accelerated radiation in their shape and form, these fossils revealed that the overall rise of flowering plants was far more gradual than Darwin had thought, and that, like all life on Earth, they had evolved their structures through a process of gradual change.

From the time when these first flowers became immortalised in the coals of the Early Cretaceous, the angiosperms – as all flowering plants are collectively known today – have since diversified into a huge range of specialised species, each one with its own way of encouraging its pollinators to disperse its pollen. Fast-growing, compared to the ancient cycads and conifers, and able to tolerate fluctuating climates, flowering plants soon became the most species-rich plant group on Earth. Relatives of the earliest flowers to evolve can still be found today, the oldest of which is a plant from the cloud forests of New Caledonia called *Amborella*, and *Nuphar*, a water-lily. The first pollinators of early angiosperms are thought to have been flying insects like scorpion flies that, having been partial to the nectar of seed ferns, would have been easily lured by the blooms of the first flowers that emerged. Angiosperms fast began to optimise their blooms to make them more enticing to particular kinds of pollinators. Over time flowers became bigger, brighter and more scented, and as flowers began to evolve to favour the tastes and temptations of certain animals, the pollinators in turn began to evolve to maximise their ability to drink nectar or eat the nutritious starchy pollen of particular flowers.

One particularly ingenious example of a flowering plant which has maximised its pollen-spreading success is the flowering tree *Gustavia longifolia*, from the western Amazonian forests. A team of tropical horticulturalists who studied the plant at the Royal Botanic Gardens at Kew found that its fleshy, deep-purple flowers have acquired a very clever method to ensure that its pollen is only taken by the particular kinds of bee that are likely to spread that pollen to other *G. longifolia* flowers. A species of night-flying bee climbs in among the stamens of the flower to feed from the nectar inside, and as the bee drinks from the sugary fluid the frequency of its buzz causes the flower to shake violently, at a force calculated to be as much as 30 times the pull of Earth's gravity. These violent vibrations shake the flower's anthers, and in the process its sticky yellow pollen is released and showered over the bee's back. This clever mechanism, called buzz-pollination, occurs in a number of other unrelated flowers from all over the world. The flowers of the tomato plant, for instance, release their pollen for only a handful of species of bee. Farmers who grow acres of the plants have tried to trick the flowers into releasing their pollen by using vibrating tuning forks or buzzing electric toothbrushes – but nothing provides as good a pollination service as the bumble bee that has evolved alongside the tomato plant for millions of years.

Flowering plants are one of the most successful life forms on the planet, and they have come to occupy almost every known habitat on Earth. But while bees are indeed the most prolific pollinators, there are many other species which help move pollen from one flower to another. Some of the plants alive today, whose ancestors evolved shortly after the primitive flowers of the Early Cretaceous, such as the water-lilies and the magnolias, evolved a wealth of tactics to persuade flies and ancient beetles to feed and transfer their pollen. As well

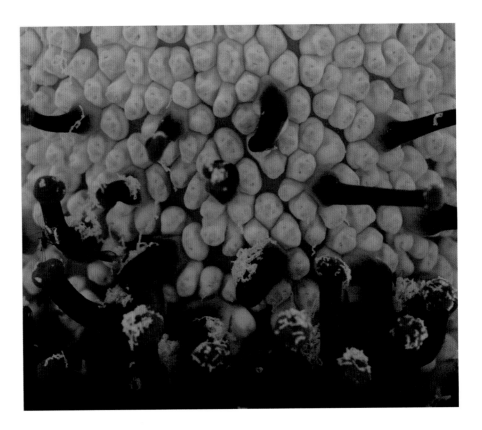

above: Flowering
giant

Structures at the base of the titan
arum's spathe produce the pollen
which is picked up by visiting insects.

—

as their heady scents and enticing blooms of electric blue, loud pinks and mesmerising yellows which make them irresistible to insects, they are also able to produce heat. This ability, known as thermogenesis, provides a warm landing place with a ready supply of nutritious pollen, and a flower is therefore an easy choice for any insect looking for an inviting place to visit.

From the bird world, hummingbirds are prolific pollinators of bright red jungle plants such as honeysuckle, using their long beaks to gather nectar from the trumpet-shaped flowers. Moths pollinate some of the more ghostly flowers, such as those of the night-blooming cacti *Echinopsis* and *Selenicereus*, and butterflies are responsible for pollinating many thousands of species of pink or lavender-coloured tropical flowers such as the Asian buddleja or the American passion-vine. Snails and slugs smear pollen from plant to plant as they move through vegetation, and mosquitoes pollinate some species of orchids. Mammals too, both on the ground and in the air, transfer pollen for many hundreds of different flowers. Even lizards on the island of Mauritius have been found to transport the sticky pollen of particular plants with tough flowers, as they forage for fruits.

Since the arrival of flowers, the animal world has been inextricably linked with the plant world, and for the past 140 million years they have evolved together. In most cases it is a mutually beneficial relationship, in that the animal gets a meal, and the plant spreads its DNA. The plant world's ability to harness the hungry nature of animals and get them to carry their pollen is the greatest trait that plants have acquired, and as long as there are animals waiting to get a meal, flowering plants will remain the dominant and most fascinating organisms on the Earth.

Mankind's obsession with flowers has not waned since Victorian times, and although orchids are still the chosen obsession of many, thousands of different flowering species are now cultivated and admired in gardens and conservatories around the world. Our continued love of flowers is personified today in the beds of the Royal Botanic Gardens at Kew, now a UNESCO World Heritage Site, where flowering plants from all corners of the globe attract over two million people every year to marvel at the variety of the plant world. The Palm House is home to exotic tropical plants from jungles all over the world, and the grand structure of the Temperate House contains thousands of temperate and cool-zone plants from Asia, Australia, Africa and the Americas. But it is in one of the more recent additions to Kew's landscape that the gardens' biggest draw can be found. In the Princess of Wales Conservatory, the most technologically advanced greenhouse in the world, there is a plant that since its discovery in Asia in 1876 has not ceased to fascinate all who see it: the titan arum (*Amorphophallus titanum*).

The plant was first discovered by an Italian botanist called Odoardo Beccari, who stumbled across it during his expeditions in the tropics of Sumatra. He packaged up some of its seeds and hastily sent them back to Europe, and when a handful of these germinated a young plant eventually made its way to Kew. For over 10 years the plant grew in size at Kew, putting out mighty leaves, the size of a small tree, until in 1889 it finally produced its first flower. Amazingly, the single triffid-like bloom which emerged from this almighty plant was as large as its 2-metre-tall leaves. It seemed clear that the titan arum must surely be the largest flower in the world. However, on closer inspection it was found that its great totem-pole-like structure was actually made up of many thousands of minuscule flowers, making it by definition an inflorescence and not a single flower. Instead the accolade of the largest single flower is held by the metre-wide parasitic species *Rafflesia arnoldii*, which also grows in the tropical forests of Sumatra, and across Southeast Asia. Nonetheless, the titan arum is a botanical giant, and rising to 3 metres tall in its mightiest specimens, its bloom towers over any human. Its flower body consists of a frilled purple collar around a tall speckled-green and cream-coloured flower-bearing spike, which is made of separate male and female flowers.

Only staying open for a couple of days, the titan arum has to attract as many of its pollinators as quickly as it can, and it does this by emitting a foul and fetid stench, described as resembling rotting flesh, sour dairy and burnt sugar, which is produced by sulphur-containing compounds on its spike. The stench of the recently opened flower is in fact so vile that the artist who came to draw the first specimen that came into flower at Kew was made ill after inhaling it for too long. Shortly after opening, the base of the flowering spike begins to generate heat of around 36°C, which creates a convection current to help waft its rancid smell through the night air. By burning reserves of stored carbohydrates, the plant produces heat in waves over a few hours, and the resulting pulse of scent that is emitted acts to punch through the layer of cooler air that forms below the forest canopy. Once through this layer, the foul scent is able to travel great distances through the forest and reach the olfactory organs of its pollinators. Any carrion-beetles or flesh-flies that catch a whiff of its odour will then hungrily fly to the flower expecting to find a meal of decaying meat, and on arrival will bump into the flower spike. Cunningly, the male and female parts of the titan arum open on separate nights to prevent the flower from self-pollinating, so it has female parts at the bottom which open on the first night, and male parts on the top which open later. As flies search the flower for the source of the rancid smell they become caught in the deep cone of the flower's collar, and in order to escape they must crawl up the flower spike, getting coated in pollen as they go. They then fly to another open flower and, starting at the bottom again, crawl upwards – and in doing so cross-pollinate the flowers. Since the 1800s titan arum has flowered numerous times at botanical gardens across the globe where it is now showcased, perhaps most notably at Kew in 1926, when police were called in to control the

huge crowds that had gathered to see and smell the much-talked-about spectacle. Titan arum plants are now grown by botanical institutions and private collectors all around the world, but the occasions when their blooms emerge still make the headlines.

The dizzying diversity of flowering plants today is truly staggering, and we continue to discover further members of these incredible plants, such as orchids which only flower under the cover of darkness and palm trees which flower themselves to death, flowers which imitate an incredible array of insects, and even flowers which mimic other flowers. In their long history, spanning over 400 million years, plants have developed many amazing strategies to better survive and reproduce, and the evolution of the flower is surely one of their greatest. Not only has it allowed angiosperms to outnumber their fern and conifer ancestors by 20 to 1, it has also helped forge the relationship between humans and plants. Many of the plants that today support the human population, such as the grasses that provide cereals and sugar, the many fruits and vegetables we eat, as well as cotton, coffee and chocolate, and trees that provide building materials, are the result of the evolutionary success of these flowering plants.

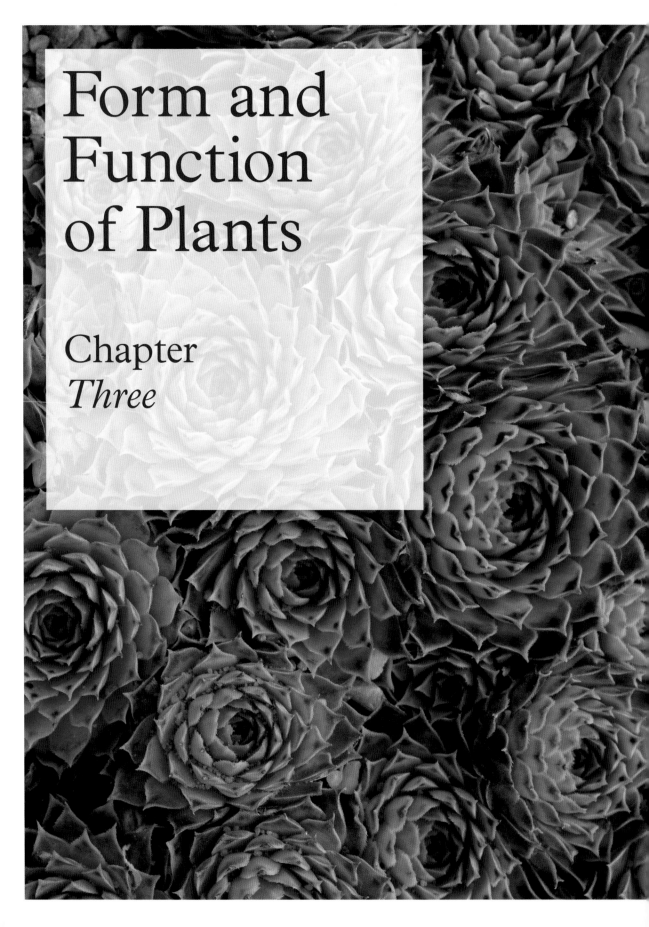

Form and Function of Plants

Chapter *Three*

'Throughout history humans have looked for patterns in nature.'

L ong before humans built the first houses out of mud, straw, wood and vines, plants employed organic material to create a multitude of structures, each one more advanced than the last. Plants use the powers of speed and size to push out the competition and overwhelm their adversaries. On the exposed slopes of wind-lashed mountainsides, plants use the powers of endurance and timing to survive the bitter winters. And in the bleakest deserts, plants use extreme structures to protect their bodies and employ chemical tactics to protect them from enemies.

The species that we can see in the world around us today bear the scars of their evolutionary history. They are the ones who passed on their genes more successfully. Every thorn on their stems, every ridge on their leaves and every berry on their branches is the accumulated result of millions of years of evolution. For every extreme of shape and structure that appears in the plant world, there is a story of adaptation and survival, to a particular climate or lifestyle that can explain them.

Every major habitat on Earth provides conditions favourable to a particular set of plants, which is why you will never find a drought-tolerant saguaro cactus growing in the middle of a moist rainforest, and you will never find a tropical mahogany tree setting down its roots in a desert. Their 450-million-year evolutionary journey has given each species of terrestrial plant a unique set of tools for survival in its specific environment, be it in the forests, the grasslands or the deserts. In each different habitat the temperature, the amount of water, the availability of light and the terrain itself determine which groups of plants can thrive there. In turn the plants and animals which already live in a habitat can limit or facilitate the addition of other species. In this way, complex ecosystems are built around an intricate network of plant, animal and fungal life, where each species is reliant on those species below it in the food chain, and in turn gives life to those above it.

Forested habitats provide an environment most similar to those in which plant life first emerged during the Devonian explosion, and it is in these habitats that we see the greatest diversity of life today. Rainforests alone contain more than 50 per cent of the world's plant and animal species, and collectively, tropical rainforests, boreal forests and temperate forests make up 30 per cent of the Earth's landmass. This makes them the most important habitat type in terms of the carbon-capturing and oxygenating services that they provide our biosphere with. Historically forests have been a lifeline, harbouring the survivors of the ice age that hit Earth at the end of the Cambrian. Plants in this habitat managed to survive, while those elsewhere perished in the cold, dry climate. These forests now make up the oldest continuously growing habitats on Earth, at around 135 million years old. In comparison, the tropical expanse of the Amazon is a relative infant at only 40 million years old. The ancient forests that sheltered the DNA of plant life on this planet can still be visited today, in places like the Daintree Rainforest on the northeastern coast of Queensland in Australia.

In modern-day forests trees dominate the landscape, and their layers of leaves and branches create niches where many other forms of life can grow. The tallest trees benefit from broad canopies which can stretch above all other plants to absorb the most light, while plants living below have to adapt to make best use of the limited light that filters through to the lower layers. However, the trees at the top of the canopy must also endure the hottest temperatures

and lowest humidity. Some of the largest forest trees are the largest plants on the planet: for example the giant sequoia, *Sequoiadendron giganteum*, which grows in the evergreen forests of central California. The largest can reach over 80 metres high, and contain enough wood to build over 40 small houses. One tree can support as many as a hundred other species of plant and animal. Woody vines called lianas use the tall trunks of trees to creep up towards the light at the top of the canopy. As they rely completely on the support of other vegetation they invest little energy in structural support, and as a result they put all of their resources into rapid growth and leaf production. There are over 2500 species of lianas which grow in the tropical forests of Africa, Asia and the Americas. As they grow, these tiny vines may remain thin and cling to the sides of trees, or may ultimately become colossally thick stems which appear as tough as tree trunks themselves. In some forests lianas have been found to make up over 40 per cent of the total leaves in the canopy.

above: Ancient refugia

Australia's Daintree Rainforest is the oldest continually growing habitat on Earth.

—

above: Climbing vines

The vascular system of lianas has evolved to transport water hundreds of metres through the canopy.

—

Lianas take up water and nutrients for the plant from the forest floor, but in order for them to reach up to the top of the canopy they must be able to transport their nutrients through their elongated stems – as far as 900 metres in the most extreme examples. While the evolution of lianas has resulted in one of the most advanced water transport systems of any plant, there is a group of specialised plants that have found a way of living high up in the architecture of the forest trees without the need for long roots. These plants are known as epiphytes, or air plants, as they do not have roots which grow into the soil, but instead have short roots which can absorb nutrients and water from organic matter that accumulates on branches high up in the forest canopy. As these plants must absorb all of their water from the air, they can only thrive in very humid environments, and in the high-altitude montane forests of the tropics trees can become covered in epiphytic plants. Epiphytes include the mosses and lichens of temperate forests, but also a multitude of more complex plants such as orchids, ferns and some tropical cacti. However, there is one group of plants that stands out as the true masters of arboreal life, the tank bromeliads. Relatives of the pineapple, tank bromeliads live attached to the branches of trees in the rainforests of South America, where they display an amazing variety of shape and colour, from broad, green fleshy-leaved plants, to small, delicate purple and red structures. Their broad leaves are arranged into a basket-shaped rosette which acts to channel the rainwater that trickles through the forest canopy into a central reservoir, and the bases of these leaves are packed so tightly together that they are able to create a watertight tank in which the water can gather. In the largest species of tank bromeliad as much as 50 litres of water can be held between the plant's leaves, and a study conducted in Puerto Rico found that in just a 1-hectare area of rainforest as much as 50,000 litres of water can be stored, which would fill a small swimming pool.

above: Tank bromeliads

These plants provide arborial homes for a whole host of animal life.

—

The little elevated lakes that gather in the bases of bromeliads make perfect homes for a handful of species looking to escape the perils of the forest floor. Insects such as mosquitoes lay their eggs in the pools of water, and flatworms find shelter among the leaves. With this abundance of insect life comes a menagerie of larger animals that visit the pools to munch on the invertebrate feast. Inch-long salamanders come to feast in the relative safety of the plant, and in some bromeliads in Jamaica small crabs have been found to dwell, territorially defending their plant against lizards and millipedes. Tiny poison dart frogs are perhaps some of the best-known lodgers, with some species spending their entire lives, from tadpole to adult, inside the seclusion of the bromeliad's tank. A recent count of the different animals that live in the bromeliads of Ecuador found an astonishing 300 different animal species that made these plants their home. However, the bromeliad does not just provide a home for these animals to be of service, rather it accommodates them so they can provide it with food. The droppings of the frogs and salamanders, which contain the digested bodies of the insects that gather in the pool, accumulate in the water and can be absorbed by the plant as a vital source of nitrogen-rich food.

Back on the forest floor plants grow in a more conventional way, rooted to the ground. Although these plants only receive around 2 per cent of the glorious sunlight available at the top of the canopy, they have the advantage of growing directly on the nutritious layer of organic material created by an unseen army of bacteria and fungi. Species here are able to spread across the forest floor in a dense carpet of growth, employing a number of adaptations which allow them to thrive in the forest understorey. One such strategy is seen in the stripy-leafed *Tradescantia zebrina* from the forests of southern Mexico, which has purple undersides to its leaves created by a pigment called anthocyanin. As light passes through the green photosynthetic tops of the leaves the purple cells underneath act like a mirror and bounce the light back up to ensure that the maximum energy is absorbed by the plant's chlorophyll. Other plants use size to their advantage to capture as much light as is physically available. One of the most successful plants at using this strategy is the giant taro plant, *Alocasia robusta*, which has the largest undivided surface area of any leaf on the planet, growing to over 3 metres in length and over 2 metres in width. Its huge glossy leaves thrive in the understorey of the tropical forests of Asia, where they fan outwards in order to gather light throughout as much of the day as possible. Another more subtle mechanism used by plants to ensure they can gather as much light as possible is found in a small purple-leafed shade-dwelling species called *Oxalis oregana*, from the redwood forests of western USA. At the top of its short 15 cm stems it has triplets of heart-shaped green leaves which are able to move in order to track the sparse sunlight as it shines through the canopy above. However, as *Oxalis* has adapted to photosynthesise in such low levels of light, strong sunlight can actually be damaging to its cells. Consequently, should a beam of sunlight break through the canopy directly onto its leaves, in just 6 seconds the plant can tilt its leaves to a vertical angle and escape the light.

Another group of prolific plants that thrive in forests are the palms, notable for their economic importance to us. Found growing both as tall trees with mighty crowns poking through the forest canopy and in short, spiky clumps at ground level, palms can live in habitats ranging from the desert islands of the hot tropics to the milder Mediterranean climes of the subtropics. They are instantly recognisable by their distinct leaves. Resembling thick, green feathers and broad, fan-like paddles, their leaves give them a surface for absorbing energy from the sun, and their deep ridges channel rainwater away from their surfaces. Various animals also rely on palms for food and shelter: birds such as palm-nut vultures and macaws flock to the plants to eat their fleshy fruits, and small, ground-dwelling mammals like the Asian palm civet (*Paradoxurus*

hermaphroditus) forage for fruits about their bases. Madagascar has the highest diversity of palms anywhere on the planet, and because of the island's lack of herbivores the leaves of its palms lack the chemical defences and spines found elsewhere. Some of the island's palms have leaves that extend for no more than a few centimetres, but in the extreme example of the raffia palm (*Raphia farinifera*), its leaves can hang down from the crown a massive 24 metres – the longest known leaf of any plant, roughly the height of a seven-storey building. The largest seed of any known plant comes from a palm too, called the coco-de-mer or double coconut, and the largest known inflorescence comes from a species of palm called *Corypha umbraculifera*. The eighteenth-century Swedish naturalist and father of taxonomy, Carl Linnaeus, was so taken by palms that he labelled them as Principes, the order of the Princes.

Palms today provide humans with an array of useful materials and foods, and after grasses and legumes they are the most economically important plants

above: Leaf architecture

Leaves are the food-making factories of green plants, and come in a variety of shapes and sizes.

—

above: Mighty bamboo

This member of the grass family is the strongest plant on the planet.

—

on the planet. Nearly every part of the palm plant can be used for food: the sap is commonly boiled to create a sugary food called jaggery, and the oils from the flower are tapped to make a fresh drink or fermented and distilled to create a number of potent alcoholic beverages, such as the east Asian liquor arak. The sweet tips of the fresh leaf growth make a sweet salad, and the starch from the fibres of the trunk can be harvested to make a nutritious food called sago. As well as their multitude of edible uses, palms provide a variety of practical materials, both locally to where they are grown and across the world. Their wood is used to make buildings and furniture, as well as ropes, clothing and fibres, and the oil from their fruits can be turned into waxes, fuel and cheap cooking oil. Sadly, the oil produced from palms is in such high demand that great swathes of diverse tropical rainforest are being felled in Southeast Asia to be replaced with vast seas of oil-palm plantations.

Less biologically rich than their tropical counterparts, the temperate forests are characterised by leafy deciduous vegetation and conifer-covered mountains. Unlike the tropics, which often have both wet and dry seasons, temperate regions have four distinct seasons of varying warmth and precipitation, and through this cycle some of nature's most striking landscapes are created. In the mountain region of south-central China some of the most dramatic habitats of temperate vegetation can be found, nourished by the waters of the five great rivers of Asia: the Mekong, Irrawaddy, Yellow, Yangtze and Salween. These are the most biodiverse temperate habitats, and among the mountain woodlands giant pandas feed on one of the most important plants of these forests – bamboo. This mighty plant makes up a key part of the understorey of the temperate broadleaf deciduous forests.

Bamboo is a member of the grass family Poaceae. In fact, bamboo is the largest grass in the world. Separating it from all the other grasses is its tough, woody stem. This gives the strongest shoots the strength of mild steel (able to withstand around 52,000 pounds per square inch, a pressure that could crush stone), making it the strongest plant on Earth. Shooting upwards like an extending telescope, each new section of the plant extends from the centre of the old sections, and the fastest species are able to advance towards the light at a staggering rate of over 5 centimetres per hour. This amazing growing capability makes bamboo a crucial plant in its forest habitat, acting as an unrivalled soil erosion control agent. Bamboo is particularly successful at re-colonising areas of land that have previously been cleared for agriculture or cattle grazing, and the re-greening of an area of land by bamboo can help return structure and life to the forest environment.

Like palms, bamboo provides us with a vast variety of building materials and foods. Bamboo-related industries are estimated to provide a livelihood for around 1.5 billion people worldwide, making it a plant of great economic importance. In Asia it is used to create high rigs of scaffolding, some over 100 metres tall, and in Central America an area of farmed bamboo forest of just 60 hectares can provide enough material to build around a thousand small houses. But it is actually the less impressive relatives of the grass family that provide humans with an even greater service. These are the grasses that give us wheat, corn, rice and maize, and that feed the animals which give us meat, leather and wool. They are the most economically important plants on the planet, and their exploitation has shaped the face of the world as we know it.

In their wild form grasses grow primarily in the semi-arid grasslands and savannahs that make up about 20 per cent of the Earth's surface. In

above: Wild grasslands

Africa's vast protected national parks
provide a snapshot of how Earth's
ancient plains would have looked.

—

the past, thousands of grazing animals such as elk, wild
horse and saiga antelope inhabited these vast grasslands.
Today these wild habitats are much quieter. Hunters who
roamed the grasslands some 20,000 years ago caused the
mega-herbivores that once dominated the landscape to disappear. Today it
is primarily livestock that inhabit these landscapes. Protected habitats such
as Yellowstone National Park in the USA and the Masai Mara National
Reserve in Kenya now form some of the few grassland areas where one can

get a glimpse of how our grassland ecosystems would have looked without human interference.

Plants that grow in grasslands are highly resilient to drought, and are often able to withstand months without rainfall. Short hairy grasses such as june grass (*Koeleria cristata*), which grows in North America's dry prairies, have shallow roots that sit just under the surface of the ground, so that as rainwater soaks into the ground it can be absorbed by the plant. Conversely, taller grasses like elephant grass (*Pennisetum purpureum*), which grows to around 5 metres and, as its name suggests, is a favoured food of Africa's largest herbivore, have hair-like, branching roots reaching up to 6 metres in length, enabling them to reach sources of water held deeper in the soil. As a means of saving water, many grasses also have long, narrow leaves which lose far less water in the drying sun than larger leaves would.

About 50 per cent of the plants that live in grassland habitat are actual grasses, with the other 50 per cent made up of what are collectively called 'forbs' as well as occasional species of tree. Forbs are a mixture of wildflowers and broad-leafed herbs. Plants such as the prairie coneflower (*Ratibida* spp.) convert the terrain when they bloom in midsummer, with each plant producing over a hundred flowers, and the metre-high flower stalks of lupins (*Lupinus* spp.) grow along the banks of streams, erupting into spearhead-shaped blooms of white and purple. The forbs that grow between the grasses generally have tall, upright body shapes. This enables them to compete for access to light, and by growing in clumps they help support and protect one another from winds.

The vegetation of the Russian steppe or the rocky plateaux of Tibet's high-altitude grasslands would appear to be largely barren, with few distinguishing features to the landscape apart from the odd tree or rocky outcrop. Unlike tropical forests, which have a very thin layer of nutritious soil, and therefore retain most of their nutrients in the vegetation above ground, the main stores of food and energy in the grassland habitat are held underground. The nutrients are retained in the subterranean parts of the plant, locked inside specialised storage organs called bulbs, tubers and roots. As the bodies of the plants above the ground die back every year, the carbohydrates and water accumulated during their growing season are taken down into these storage organs. Deep under the ground these tubers and bulbs are further enriched by organic matter produced by decomposing plant bodies that are broken down by fungi and bacteria in the soil. In the tropical savannahs of Tanzania the scorching hot weather of the dry summer can last for over 7 months, turning the landscape into a parched, brown wilderness. During this period many of the plants must lie dormant until the rains return. At the other extreme, the winters of the Russian steppe are so bitterly cold that nothing can grow through the snow and frost, with plants only beginning to grow again when the temperatures rise above 10°C, thawing out the frozen ground.

In the savannahs of northern Australia the annual cycle of plant growth and dormancy is largely dictated by the periodic wildfires that spread through the dry summer vegetation. Far from being destructive, these bushfires are a key feature of this habitat. Triggered by lightning or from the sparks from falling rocks, these fires provide a vital service in renewing the landscape. They clear the tinderbox of dead plants left at the end of each dry season, and free up space and nutrients for new plants to grow in their place. For many millions of years these fires have periodically swept across dry grasslands, and many of the plants which thrive in those habitats have come to rely on the fires for parts of their life cycle. The paperbark tree (*Melaleuca quinquenervia*), for instance, native to coastal eastern Australia, has little buds under its thin flaky bark which cannot sprout without the intense heat of the bushfires. The eucalyptus tree, which stores up energy in a swollen root under the ground, can spring up new shoots following a fire. Some plants have even evolved reproductive strategies that rely on intermittent fires, such as the orange banksia (*Banksia prionotes*). Because of its height (up to 10 metres) and minimal foliage it is able to survive the flames of a bushfire, and as the heat around its base increases to a sweltering 265°C, its large seed-storing cones are baked. As the fire subsides and the cones cool they open up, allowing their seeds to fall to the parched ground, where they germinate.

Numerous other plants in fire-prone habitats possess seeds adapted to the occurrence of wildfires. As small, freshly germinated shoots would be instantly incinerated by the immense heat of a bushfire, the best time for seedlings to emerge is immediately after fire, when the habitat has been cleared of the most flammable vegetation, making it unlikely for a fire to take hold again for a number of months or even years. As a result, the seeds of the Californian lilac (*Ceanothus* spp.) will not germinate unless they are exposed to extreme heat, and the seeds

of the Californian flowering herb whispering bells (*Emmenanthe penduliflora*) will lie dormant until they are engulfed in the smoke of the charred vegetation released during a fire. One of the most extreme adaptations of plants to fire is seen in those that are

able to help start them. These 'plant arsonists' include members of the rockrose (*Cistus* spp.) family that characterise the dry scrublands of Morocco, Portugal and the Middle East. These short shrubby plants have a sticky aromatic leaf resin which is highly volatile, and for the most flammable of the species a dry heat of just over 32°C can be enough to cause the resin to burst into flames. Although the fire will destroy the small shrub, its seeds have a hard coat which protects them from the heat, and when the fire has burned itself out these seeds have a ready-made space in which to germinate. When it next rains, water will mix with the charcoaled ground and the chemicals left by the burning dissolve in the water and coat the seeds, triggering them to put out their first roots and shoots into the baked soil.

Perhaps the most robust seed of any plant in the wild is that of the aquatic perennial *Nelumbo nucifera* or sacred lotus, which grows in the wet grasslands of Asia, the Middle East and Australia. Fossil evidence for its morphologically similar cousins indicates that these plants were growing during the Early Cretaceous period – between 145 to 100 million years ago – making this plant one of the oldest flowering plants on the planet. The secret to its survival comes from its ability to put out new shoots from submerged root tubers called rhizomes, and in this way one individual plant can grow to cover an entire lake in just a matter of months. With its rhizomes stuck deep into the mud, the plant puts out distinct, round leaves, which sit up out of the water on tall thick stems. Once a year vast seas of prolific *Nelumbo* burst into flower with scented, 20-centimetre-wide blooms of pink, red and white, which open in the morning and close up again at night. Following pollination, they produce marble-like seeds, about the size of an olive, from a large central cone structure, with each seed sitting in its own pocket at the top of the cone.

The *Nelumbo*'s seeds, which botanically are actually nutlets borne from a multiple fruit, are covered with an amazingly hard, woody coating, which is almost completely impermeable to water. After falling from the plant the rock-hard seeds sink to the bottom of their watery habitat, where they can remain for hundreds of years. Some seeds that were retrieved from an ancient peat bog in Manchuria were dated as over 1000 years old, and after being exposed to water were still able to germinate. In an attempt to test the true strength of these seeds, they have been subjected to blowtorch flames, buried in concrete, and bashed with hammers, all with no detrimental effect whatsoever. Though all seeds require water to germinate, the *Nelumbo* has developed a unique survival strategy. First, as the leaves of the plant quickly swamp any area where it grows, every bit of growing space is taken, and if a seed did sprout there it would struggle to survive in the shade of the adult plants. Secondly, the tubers at the roots of the *Nelumbo* are eaten by grazing animals such as muskrats and beavers, and families of these animals often move to the *Nelumbo*'s habitat, where they ravenously feast on the plants, increasing in numbers until the whole area has been cleared of any *Nelumbo*. If the seeds had produced new plants during this time they too would have been eaten. Therefore it is advantageous for the

opposite: Lotus effect

The hydrophobic surface of the *Nelumbo* leaf has been the inspiration of a number of nanotechnologies.

—

seeds to remain dormant until the herbivores have moved on, after which they can sprout and re-colonise the area again.

Almost as amazing as the endurance of the *Nelumbo* seed is the speed at which it can sprout once its inner layers are exposed to water. In the bottom of dry lake beds the tough seed coating will get scraped and scratched during its long dormancy, and over many years, millimetre by millimetre, its tough outer shell can be worn away. At one end of the seed is a small dimple, and as soon as the softer layer underneath becomes exposed water is able to seep in. In just 24 hours it can almost double in size. After another day the seed begins to crack open, and extending out from the cracked coating like a miniature green arm, the tiny plant that has lain dormant inside the seed for many years emerges, unfolding to form a 10-centimetre-long spike. Even if only one seed from a stock of hundreds of thousands manages to break its dormancy and germinate, that one plant could potentially colonise many square kilometres of wetland.

As well as the longevity of its seed and the unstoppable growth of its green parts, the *Nelumbo*'s leaves have helped it become such a prolific survivor in its aquatic habitat. Perched horizontally on top of their tall stems, the round green leaves of the *Nelumbo* closely resemble the water-lily pads of their aquatic neighbours. However, it is not until it begins to rain that the incredible properties of the *Nelumbo*'s leaves reveal themselves. As drops of water fall onto the pads of water-lilies they simply splash and spread over the leaf, and as drops of rain splash into the murky lake water, this too splashes the leaf, showering it in muddy water. But the drops of water that fall on the surface of the *Nelumbo* leaves appear to bounce and bead. The water simply rolls off the leaf, leaving it completely dry. What is more, any flecks of muddy lake water that could potentially hinder its ability to photosynthesise are gathered up in these beads of rainwater, which clean the surface of the leaf. This 'self-cleaning' effect of the leaves has been known in Asia for over 2000 years, even gaining a mention in the sacred Hindu text the *Bhagavad Gita*. However, it was the invention of the scanning electron microscope in the early 1970s which helped scientists to discover the secret abilities/traits of the *Nelumbo* leaf. For the first time scientists were able to see the microscopic details of these leaves. Instead of having a smooth surface, it was found that the leaf is covered with a nanoscopic layer of bumpy cells, with each ridge and furrow coated in a thin layer of wax crystals which repel

the water. As water drops land on this surface they are formed into perfectly round beads, which can only contact the leaf with as much as 3 per cent of their surface. Unable to spread out and wet the surface, they simply glide off. Scientists coined the term the 'lotus effect' to describe the amazing properties of this type of structure. Nowadays the patented 'lotus effect' is used in technologies as wide-ranging as anti-fog microchips, water-repellent windows and solar panels, and it has even been used to create self-cleaning fabrics.

above: Prolific lotus

The ability for *Nelumbo nucifera* to put out new shoots from root tubers enables it to colonise huge areas of wetland.

———

At the very core of all life is DNA, itself a meticulous geometric shape, and emanating outwards from this precise core, all plant life grows in a strict pattern. The layers of light-capturing leaves that make up the forest vegetation may seem to shoot and sprout at any angle possible. However, what appears at ground level to be an erratic budding of leaves reveals itself from above to be a clever geometric stacking of light-absorbing surfaces, set out in a deliberate mosaic so that maximum light can be absorbed by chlorophyll at all times of the day. Coils of leaves arranged around the stems of *Costus* take the shape of miniature spiral staircases, resembling macro-stacks of helical green DNA, allowing the maximum height between overlapping leaves in order to reduce self-shading. Some trees, such as the umbrella thorn (*Acacia tortilis*), arrange their leaves into a monolayer of green to create a single interlocking patchwork canopy. Other trees, for example the English oak (*Quercus robur*), orientate their leaves into multiple layers, with smaller ribbed leaves at their tops to allow light to filter through to the broader leaves below.

Renaissance polymath Leonardo da Vinci was fascinated by innumerable patterns he found in nature, not least those of the plant world. He took extensive notes on the subject of these geometric formations, and outlined a simple formula which he believed could determine the size and shape of all trees. Leonardo stated that 'all the branches of a tree at every stage of its height when put together are equal in thickness to the trunk.' Or, for the mathematically inclined, the squared diameter of a tree's trunk is equal to the squared sum of the diameters of its branches. Incredibly, when tested, this simple formula was found to hold true. For a long time it was assumed that trees follow this defined model of growth to ensure that their trunk and branches are always in the correct ratio to sufficiently transport sap to all of their leaves. However in 2011 a French physicist called Christophe Eloy suspected that this was not the case. Having spent many years studying fluid mechanics and the way in which air flows around solid

'Mathematical rules have been discovered which underpin many of the shapes and patterns we see in the plant world.'

above: Spiral leaves

This organisation of leaves acts to
maximise the plant's exposure
to light.

—

objects, Eloy proposed that trees did not follow Leonardo's formula to enable efficient transport through their parts, but so that trees of all sizes are able to support themselves in strong winds. By creating computer models of various trees of the optimum shape in relation to their size to withstand the force of the wind, what he found again and again was that they all matched Leonardo's rule. Amazingly, he had revealed that in the same way that engineers use wind load calculation to ensure that tall structures such as the Eiffel Tower remain stable in all weathers, trees appear to be shaped by the very same design principles, only they have been doing it for many millions of years.

Throughout history humans have looked for patterns in nature. The fronds of fern branches reveal geometric fractal formations of repeating feather-like structures, increasing in size at each level to create the whole plant. The term fractal was first coined by French-American mathematician Benoît Mandelbrot, who described it as a body possessing what he called 'self-similar' patterns of complexity, increasing with magnification, which if divided into parts gives you a nearly identical reduced-size copy of the whole. In his 1993 book entitled *Fractals Everywhere*, British mathematician Michael Barnsley explained that the shape of the fern leaf is in fact a product of chaotic mathematical equations. He produced a formula which proved that if random numbers were plotted over a long enough period of time and converted into geometric shapes, ultimately the outcome will be a branching shape with different frond-like forms at different scales, almost identical to that of a fern leaf.

The same pattern of repeated identical structures at increasing scales that gives ferns their shape is also seen in the vascular systems of leaves. Held up to the light, their dendritic patterns are revealed: tiny veins connect to branches of larger veins, which in turn create branches of larger veins, attached to the leaf's midrib. This up-scaling from small structures to create a larger whole is one of the natural world's most elegant patterns. It is reflected in the swirls of ammonite shells, the frozen arms of snowflakes, and the patterns of branching river deltas. But most importantly it is a very efficient way for a plant to program its growth. The amount of DNA required to code for the same structure repeated at different scales is far less than that needed to make numerous different structures. This system provides a highly economical pattern of growth. Perhaps the most extreme example of plant fractal growth can be seen in the structure of the Romanesco broccoli, and the branches of many trees exhibit a similar shape of growth, albeit less dramatic.

Mathematical rules have been discovered which underpin many of the shapes and patterns we see in the plant world, and specific sequences of numbers have been uncovered which can determine the number of petals a plant will have, how its seeds are arranged in its head, or where its leaves sprout along its stem. If you were to count the number of florets on the herbaceous chicory you will find that they always number 21, while those of daisies usually number 34 or 55. Lilies have three petals and buttercup flowers always have five petals, and you will not find many flowers with four petals. Take a look at a pinecone and you will discover two sets of spirals, eight from one side and 13 from the other, while others have five from one side and eight from the other. These seemingly random numbers, far from being accidental, are recognised to be part of what is known as the Fibonacci sequence – a sequence in which each number is created by the sum of the two preceding it (e.g. 1, 2, 3, 5, 8, 13, ad infinitum). These numbers repeatedly occur in the plant kingdom. If you cut open the fruit of many flowering plants the number of segments is governed by Fibonacci numbers, and the number of branches at each level of a plant will often follow a pattern of the Fibonacci sequence.

This unique set of numbers is woven throughout the very fabric of the shapes and patterns that we see around us, but there is another even more fascinating way in which these numbers have shaped the natural world, and that is from a number which is derived from the Fibonacci sequence, called the Golden Ratio. If you take each number of the sequence and divide it by the previous number, (e.g. 2/1, 3/2, 5/3, 8/5) you will soon see a pattern emerge of similar answers like 1.6666, 1.6 and 1.625, and as you continue along the sequence you will eventually hit upon the Golden Ratio, which is approximately 1.618. From controlling the way in which growing plants stack their cells, to determining how plants orientate their leaves on their stems, this Golden Ratio is the mathematical blueprint that shapes much of the plant world around us. As a plant begins to grow from a group of cells, these begin to form a natural spiral formation, with each new cell emerging after a rotation. How much each emerging cell turns to create the optimum structure is determined by the Golden Ratio, a rotation of 0.6 (which would be the same as rotating 1.6). This pattern of growth is equally mirrored in each individual cell and in the large fleshy structures of some plants. The head of a sunflower crossed with a pattern of tightly packed achenes (dry fruit contained within a shell) is one of the best examples of this tight spiralling formation, with each kernel growing next to its neighbour at an angle corresponding exactly to the Golden Ratio. It is visible also in the hexagonal segments which make up the flesh of a pineapple fruit, in the spiralling leaves of many agave plants, and in the flowering heads of daisies. The patterns created by these numerical patterns extend far beyond the plant world, and can be seen in marine shells, hurricanes, and even distant galaxies. Through these shared patterns we see an intrinsic unity within nature, as the same fundamental blueprints that give structure to the cosmos give shape to the plant life of our planet.

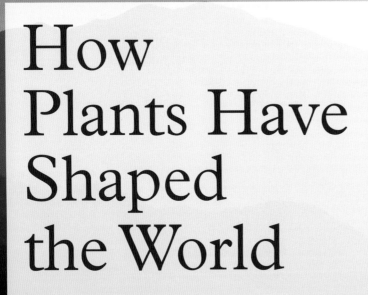

How Plants Have Shaped the World

Chapter
Four

'It is imperative that the breadth of plant diversity on Earth is maintained for its ecological worth alone.'

The diversity of plant life on Earth today provides our human species with two of the most vital ingredients that we require for survival: food and oxygen. We have come to rely on an incredible array of resources from plants, such as building materials, medicines, clothes and cosmetics. They have influenced the very geology and chemical composition of the world that we inhabit today. Over the past 500 million years the oxygen released into the air by photosynthesising organisms has shaped Earth's climate, and so plants have ultimately shaped the path of life itself. On a global scale today, the tropical forests and grasslands help control the cycling of carbon dioxide, they affect the rate at which rocks erode and landscapes shift, and they alter the amount of light that is reflected or absorbed from the surface of the Earth, thus changing the temperature of the environment.

Since their emergence some 500 million years ago, plants have had an influence on everything around them. Long before the Incas created their civilisation in the jungles of Latin America, long before the hunter-gatherers of the Euphrates planted the first seeds of wheat 13,000 years ago, and long before the early human architects gave shape to their first stone dwellings, plants had already carved their path across the face of the planet, staking their claim as a dynamic and integral part of Earth. It is only recently, however, that we have been able to unravel the impact that plants have had on shaping the history of our planet.

Oxygen

Oxygen supports 99 per cent of life on our planet. Today this elemental gas makes up 21 per cent of our atmosphere, a comparably small amount compared with nitrogen, which makes up 78 per cent. The oxygen in our atmosphere today has built up because of the synthetic activity of green plants, and is essential for the existence of animals and most other living organisms on our planet. Until humans are able to invent a way of synthesising oxygen and turning sunlight and simple molecules into food, we will continue to rely on plants for every facet of our lives, from the air we breathe and the clothes we wear, to the food we eat and the medicines we use.

When oxygen first appeared in the oceans over two billion years ago it was toxic to life. Oxygen rapidly devastates organic compounds, and as a result its presence in the ancient oceans killed off groups of primitive seabed organisms that could not adapt. However, as the oxygenation of the planet occurred over many hundreds of millions of years, large numbers of organisms were able to evolve which could tolerate and make good use of the increasing levels of oxygen. It was the photosynthetic ability of green algae around 500 million years ago that initially pumped significant amounts of free oxygen into the atmosphere and first tipped the balance of asphyxiating gases towards a more accommodating atmosphere, and this very same chemical process of photosynthesis is still at work today. Thanks to the activity of the green parts of plants all over the world, sucking up carbon dioxide like a sponge and releasing oxygen, life in its myriad forms remains viable.

above: Harnessing light

Thin leaves packed with chlorophyll allow plants to absorb as much energy as possible from the sun.

opposite: Wonder molecule

The oxygen produced by plants is vital to all aerobic organisms.

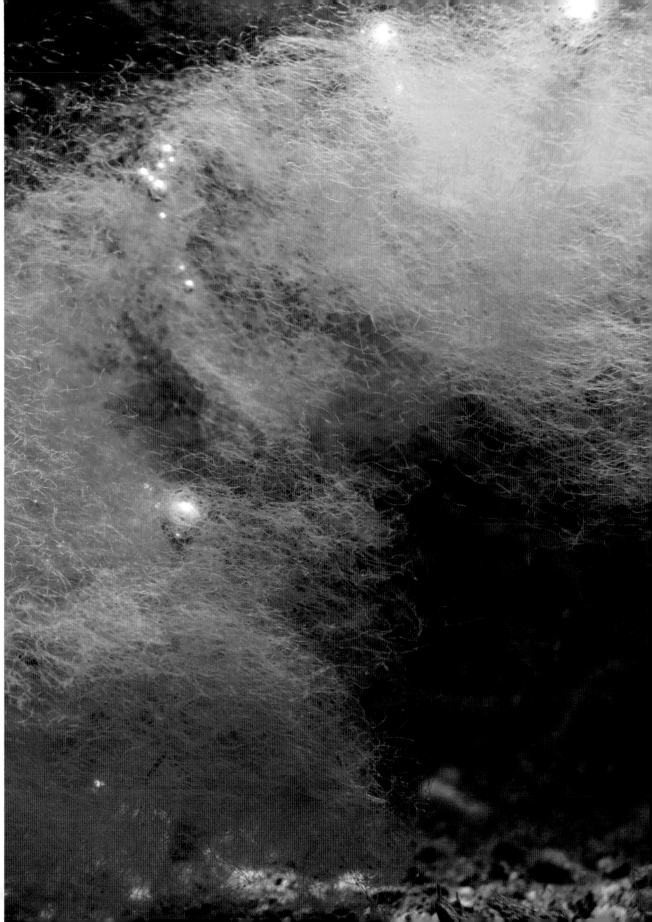

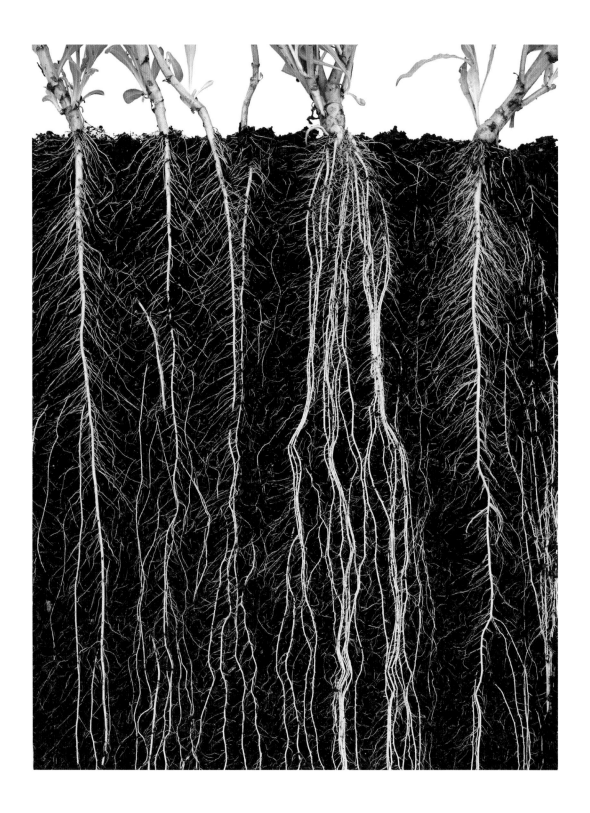

KINGDOM OF PLANTS

opposite: Root power

Dense underground networks of roots not only help enrich the soil but prevent it from being eroded away.
—

The first leafless land plants would have had a fairly small impact on the weathering of the landscape, but as terrestrial plants with leaves increased in size and diversity, their ability to lock away carbon dioxide would have increased greatly. As carbon dioxide levels dropped and oxygen levels increased, plants would have needed to increase the density of the gas-exchanging structures on their leaves, called stomata, in order to maintain their rate of photosynthesis in the face of dwindling carbon dioxide levels. With less carbon dioxide in the air, larger leaves with more stomata would have been at an advantage, and this is believed to have been the primary driver for plants to grow larger leaves. Previously a plant with large leaves would have been prone to drying out, but the reverse greenhouse effect that was being caused by the removal of carbon dioxide from the atmosphere had provided Earth with a much cooler climate. As the protective layer of greenhouse gas was slowly absorbed into the ground more of the sun's heat was able to reflect back out of the atmosphere, and by around 300 million years ago the average surface temperature of the planet fell to about 14°C, roughly one degree cooler than our present-day climate.

As competition for increasingly larger leaves drove the evolution of these towering forests, oxygen started to fill the atmosphere, reaching a staggering 35 per cent, compared with 21 per cent today. The fossil evidence of the animals that roamed the land at this time shows that they had a dramatic response to this abundant oxygen, and they grew like never before. Less effort was required for insects to pump oxygen to their organs and they were able to develop massive bodies, giving rise to a wealth of giant fauna – monstrous dragonflies with 75-centimetre wingspans, scorpions over half a metre in length and millipedes over one and a half metres long. However, as the green forests spread across the land, the protective greenhouse layer was reduced, sending the planet into a climate crisis. Over a few thousand years the temperature dramatically dropped. Ice caps began to form in the southern hemisphere, which in turn drew moisture out of the atmosphere, making the air dry and inhospitable for many plants and animals. These polar ice sheets increased the amount of sunlight reflected back into space and a perilous positive feedback loop was created, whereby more ice created more cooling, which in turn produced more ice. The Earth was enter-

above: Snowball Earth

The formation of ice caps during the Karoo Ice Age caused the Earth to become cooler and drier.

—

ing what would later be known as the Karoo Ice Age, a mighty freeze which would grip the planet for a hundred million years.

As the Earth spiralled headlong into a deep ice age, plants and animals began to suffer in the cold arid climate and many thousands of species became extinct. Rainforests were reduced to a few isolated pockets in sheltered valleys and cool mountain tops. As photosynthesising life had been so drastically reduced by the extreme climate, only a relatively tiny proportion survived. As the number of plants dwindled, so the amount of carbon dioxide that was absorbed dropped massively. Combined with this, a long period of volcanic activity in Siberia acted to replenish further the carbon dioxide levels in the atmosphere, and once again the scales had been tipped in favour of life on Earth. The effects were not instant by any means, and rocks from the Karoo region in South Africa suggest that the ice advanced and retreated several times over many millions of years, but in time the climate stabilised and eventually the ice sheets began to shrink.

As the climate warmed once more, water released from the ice sheets bathed the land as warm moist air and swathes of green vegetation returned the landscape to its former glory. Where previously forests of tree ferns and gymnosperms had dominated terrestrial habitats, these were now replaced with extensive woodlands of conifers and cycads. From the swamps and banks of the river deltas, reptilian life forms emerged, and when the ice melted and sea levels rose, massive areas of coral reef expanded to keep up with the rising water. As heavy rains returned to the post-ice-age lands, huge deposits of sediment were washed into the tropical seas, covering the reefs. Crushed under the weight of hundreds of metres of sediment, these reefs were eventually turned into limestone, and over 200 million years later they now exist as huge tree-covered-mountain ranges across the jungles of Southeast Asia. The 100-million-year chapter which spanned a period of extreme climate change has had a fundamental impact on our planet. From shaping the diversity of life in our oceans, to accelerating the evolution of land animals, the arrival of photosynthesising green plants has changed the world.

Ecosystems

Plants continue to play an indispensable role in regulating our planet. Through photosynthesis and respiration they help maintain the necessary equilibrium of gases in the air, the powerful action of their roots recycles nutrients and stabilises the habitats that surround them, and they help regulate the global temperature by the amount of sunlight reflected back into space. Plants underpin the food chains that drive the world's major ecosystems. Without the expanse of Africa's savannahs there would be no herds of buffalo or impala, and in turn there would be no prides of lions or hyena which feast on these animals. But plants do not just help sustain these large animals, they also provide vital food and shelter for the plethora of invertebrates that make up a large proportion of our global biodiversity.

As well as pumping oxygen out of their leaves, trees play a crucial role in maintaining the equilibrium of gases in our biosphere. Each year the world's forests create around 105 billion tonnes of biomass from the carbon dioxide that they absorb from the air, storing it in their bark and trunks for many centuries. In the past, huge tracts of carbon-storing trees died and were compacted in swampy areas that, over time, formed the bands of carbon-rich coal that we now extract from the ground to use as fuel. This process continues today in areas such as the Florida Everglades and Okefenokee Swamp in Georgia, vegetation decays and subsides, and is getting covered by sediment. For the past 10,000 years the processes of photosynthesis, respiration and carbon storage have created a stable atmosphere which is beneficial to the organisms which inhabit the Earth. Alongside this, plant life also plays a key role in the planet's water cycle. As the tiny pores in leaves open to allow carbon dioxide to enter, water escapes from the plant. More water is subsequently taken in by the roots and passed up through the plant's vascular system to replace it. Through this process vast rainforests such as the Amazon take up the rainwater from torrential rainstorms and return it to the air as water vapour, providing an important role in cloud formation and helping to cycle water between the oceans, the land and the atmosphere.

The plants of the planet's wetlands also play an important role in maintaining the equilibrium of the water cycle. These habitats occur in low-lying ground alongside lakes and river systems, and are made up of many species, including grass-like sedges and tall round-stemmed rushes, as well as free-floating plants such as water-lilies, duckweed and frogbit. By holding water in the ground around their roots, these plants slow the flow of water through the ground, thus preventing flooding. Following heavy rains, a wetland area of one hectare can store well over 10 million litres of water, which it then slowly releases over time. As the water seeps through the roots and shoots of the vegetation they act like a giant filter to remove pollutants and purify the water. Covering about 6 per cent of the Earth's surface, wetland habitats also have an economic importance for the populations of people living in or around them, providing food, clean water and building materials, and preventing the erosion of the landscape. As well as inland wetlands such as the floodplains of Africa, coastal wetland habitats, such as intertidal marshes and mangrove swamps, also play a crucial role, acting as a trap for sediments and as a sink for nutrients, as well as helping to protect the coastline from erosion. The extensive root systems of mangroves maintain the long-term stability of the coasts where they grow, in the process providing food, shelter and breeding grounds for a multitude of fish and other wildlife. Global wetlands are estimated to have an economic

above: Crucial habitat

Wetlands are hugely important habitats for fish, wildlife and society.

value of around $3.4 billion, in relation to the flood protection and resources they offer.

It is imperative that the breadth of plant diversity on the planet is maintained for its ecological worth alone. The intricate web of plant and animal life that maintains the tropical rainforests, the vast grasslands and the water-filled wetlands can only exist if these areas are protected and conserved. A paper published in the journal *Nature* in 2011 assessed the importance of plant diversity in particular habitats, and found that while in one given year a habitat may be able to remain healthy with only a small percentage of its plant species, the natural fluctuations that occur in a habitat over multiple years mean that it relies on as many as 84 per cent of its species before it starts to decline in health.

The loss of a single key plant or animal species can bring an ecosystem to disintegration. An expanding human population, the introduction of non-native species and the destruction of wetlands all threaten to weaken these natural habitats, and so an understanding of the importance of these areas is crucial to safeguarding their future. Deforestation, over-grazing of livestock and intensive farming expose the soil to the air. Very quickly, the thick layer of nutrient-rich soil that has built up over many years is worn away, and the land becomes desolate and infertile. The animals that once thrived there either die or have to move into other areas, potentially upsetting the balance of other habitats. Ecologists can predict many of the knock-on effects that the loss of certain habitats could have, but many of the most devastating consequences cannot be known until it is too late.

Plants and Humans

The human race is faced with a number of major ecological and economic problems: the warming of the planet, drought, famines, widespread poverty and political instability. The causes of these problems are wide-ranging and complex, but plants have potential to help return prosperity to the planet.

The human relationship with plants has shaped communities and driven civilisations, and has also crippled nations. In the beginning, our hunter-gatherer ancestors collected fruits, roots and berries to supplement their diet of small mammals and birds, as they travelled on foot across the land. The inquisitive nature of early humans led them to discover the most nutritious plants; the chemical compounds of some made them sick, while others stimulated the body and mind. In time the seeds of favoured food plants began to spring up from the dung heaps of their hunting trails. Around 13,000 years ago the hunter-gatherers who inhabited the Euphrates Valley collected over a hundred different species of edible seeds and fruits to supplement their diet. Then around 11,000 years ago the environment changed to a much colder and drier climate, and many of the grasses and cereal plants that they had come to rely on stopped growing. In order to survive the drought the hunter-gatherers had to take the seeds of the most easily grown wild plants from the low-lying areas, and cultivate them in the moist soils of the mountain slopes. As these cereal plants, such as rye, lentils and wheat, would not have been

opposite: Birth of agriculture

The ability of humans to cultivate wild grasses and cereals led to a vast population expansion.

—

above: Power foods

Packed with vitamins and minerals, fruit and vegetables are the building blocks of any healthy diet.

overleaf: Ancient reefs

The towering mountains of Thailand's Khao Sok National Park are the remains of ancient coral reefs.

able to out-compete the natural shrubs of these areas, it was necessary for the early cultivators to clear the natural vegetation, and in doing so the first agriculturalists were born.

As a more amiable climate of warmer and wetter conditions returned to the Near East around 10,000 years ago, the farming of cereals spread through the region, and as food supplies increased and became more stable, so populations increased and it became easier for communities to settle down in larger, more permanent villages. These agriculturalists became reliant on being able to grow enough food in order to support their growing dependent populations, thus inextricably linking them to the plant world. Humans were no longer simply living in their environment, they were managing it. To increase the productivity of their crops and to receive the most nutrients from the cereals and vegetables that they grew, only the hardiest plants were selected for cultivation. By choosing only to sow the seeds from plants that had larger fruits, fatter heads or smaller inedible parts, over many generations the crops that were farmed soon began to look different from their wild counterparts. Of the 20,000 or so species of known edible plants, only around 3000 are eaten by humans, and only around 200 of these have ever been domesticated through agriculture. Today a mere 12 species of these domesticated plants make up over three-quarters of human calorie intake across the world. These 12 staple crops are: potatoes, rice, wheat, sugar cane, sorghum, soya beans, cassava, bananas, maize, millet, beans and sweet potatoes. Thanks to the plant-collecting expeditions of the Russian botanist and geneticist Nikolai Vavilov in the early 1900s, each of these modern cultivated species can be traced back to their origins. It is still possible to find some of the wild relatives of domesticated species, although many of them bear little resemblance to the plants we eat in the twenty-first century – wild maize varieties from Mexico can be as thin as a finger, wild potatoes from Peru can be red or bluish-purple, ranging from round and fat to long and thin, and the ancestors of the banana are short, straight and packed with pea-sized seeds. These ancestral plants contain a wealth of important genetic material, and by cross-breeding them with modern cultivated species, they can be used to increase the genetic vigour that has been lost through millennia of farming.

For many thousands of years, since humankind began to farm and selectively breed crops, there was no real understanding of the processes by which plants of a certain size or shape passed on their characteristics to the next generation. Ancient farmers would try and improve their crops by brushing the pollen from one species onto another to try and combine desirable traits, and sometimes they were successful. But it was not until an Austrian monk named Gregor Mendel took up the study of peas in the 1850s that anyone truly understood the nature of genetic inheritance in plants. What Mendel postulated was that every trait of a plant, such as its colour and shape, is controlled by a pair of unit factors – which we understand now as pairs of genes. For example, a flower may have a pair of unit factors which dictate its petal colour. He stated that these pairs can have both dominant and recessive forms and that when combined in an organism the dominant form presides over the recessive form for that trait. For example, if a plant contains a dominant unit factor for red petal colour and a recessive unit factor for white petal colour, the plant will grow red petals. Finally Mendel proposed that these pairs of unit factors were split randomly during the production of a plant's sex cells or gametes. From these few suppositions Mendel gave the world the first outline of the concept of the gene, and provided a valuable illumination of how desirable traits could be selectively bred for in plants.

With this knowledge, botanists and farmers were able to create hybrid plants containing the best of many different strains. Experiments in the early twentieth century by American geneticist and maize breeder Donald Jones managed to cross four different types of corn to create a super-hybrid which was high in protein and grew faster and stronger than any previous strains. This new hybrid corn soon became the desired crop of many US farmers, and between 1930 and 1980 crop yields rose from 2000 litres per hectare to over 7220. Within this period the world of agriculture also saw huge advances in farming technology and the use of pesticides, which boosted yields across the globe. This marked the age of the Green Revolution. With the current pressures of an increasing human population and increased consumption rates, it is calculated that food production rates need to triple their current rates if we are to feed the expected global population in 2050.

Sugar

By the sixteenth century the major civilisations of the world had long been master agriculturalists and had developed tools and farming techniques which enabled them to produce large quantities of food all year round. During this time, the Golden Age of Exploration saw trade routes open up between countries and kingdoms across the globe, and the traffic of exotic crops of tea, cotton, sugar, rubber and tobacco revolutionised the regions where they were grown. Plants that were commonplace in their country of origin became near-priceless overseas, and the astonishing prices that members of high society would pay to enjoy the latest pastime of pipe smoking, or to have an exotic pineapple to show off at their formal dinners, fuelled a roaring trade in plants and plant products. However, the most important plant commodity came to be sugar cane.

below: Sugar cane

Saccharum officinarum, one of six species of grass grown for their high sugar content.

—

Sugar cane belongs to a group of giant tropical grasses called *Saccharum*, native to New Guinea, which thrive in areas with long, warm growing seasons and lots of sun. All green plants produce natural sweet-tasting sugars, through photosynthesis, which they store as sugar or starch in their roots, stems, sap, seeds and fruit. But sugar cane builds up these sugars in exceptionally large quantities in its fleshy 6-metre-tall stems. The sugar cane plant has been used by humans for over 8000 years, and was originally harvested for use in roof-thatching and for chewing; after removing its hard outer layer the inner tough fibres could then be sucked and chewed to release the sugary juice from the stem. From around 1000 BC people produced crystallised sugar, which could be used to sweeten foods, by boiling the cane juice, and its cultivation gradually spread along human migration routes to Southeast Asia and India and east into the Pacific. Arab armies and traders who made their way to India in the seventh century discovered what they described as the 'reed that could produce honey without bees', and they soon returned with the plant to their lands in northern Africa to establish their own plantations of sugar cane. Arab merchants also traded sugar cane with Spain and Portugal, and because it was such a highly profitable crop those countries soon began to search for new places where it could be grown.

On his second voyage to the New World, Christopher Columbus took with him some sugar cane plants, as he believed they could grow successfully in the tropical climate of the West Indies. In 1493 the first trial crop was planted in the fertile soils of the Caribbean island of Hispaniola. The heavy rain and long hours of tropical sun were perfect for the plant to thrive, and he soon reported back to Queen Isabella of Spain that it grew faster there than anywhere else in the world. News soon spread of this untapped fertile land, and famers from Britain, France and Holland hastily headed to Brazil, Cuba, Mexico and the West Indies to start their own plantations.

Back in Europe, the demand for sugar was driven by the campaigns which touted its numerous health benefits and heralded it as a new wonder food. Very soon it became fashionable to use it as a sweetener in various drinks such as cocoa, tea and coffee. Initially the plantations of Cuba, Brazil and Mexico employed the local populations to cut and process the sugar cane crop, but as the overseas demand for sugar skyrocketed, increasing numbers of work-ers were needed to maintain the productivity of the West Indies plantations.

above: Sugar cane

This tropical grass was the engine of the slave trade, which saw millions of African slaves brought to the Americas.

opposite: White gold

Sugar makes up an important part of our modern diet, with every teaspoon containing around 16 calories.

European colonial exploits in Africa at the time had tapped into a cheap and powerful workforce in the form of hundreds of thousands of slaves, and it was not long before their potential on the plantations was realised. Ships soon left Britain laden with goods destined for West Africa, where a cruel trade took place exchanging cargo for slaves, who would then be taken to the West Indies to toil on the plantations. This became known as 'Triangular Trade', and by the 1790s these trade routes saw a staggering 12 million African slaves transported in harrowing conditions to the colonies of the western hemisphere, with many of them perishing along the way. For around two hundred years the booming sugar industry thrived on a workforce of slaves, and soon every island of the Caribbean was covered with plantations and mills for refining the cane. Between the 1600s and 1800s, this tropical grass drove the economies of Europe, the Americas, Asia and Africa. Eventually the successful slave revolts which began in the French colony of Saint Domingue in 1791 heralded the beginning of the end of the regime of slavery. Nonetheless, the complete abolition of slavery was only completed a century later when the slaves of Cuba and Brazil finally gained their freedom. The two-hundred-year shipping of Africans across the Atlantic to supply the sugar industry of the Caribbean and Americas created such a massive diaspora that its effects are still felt by many millions of people today.

Until 1793 sugar cane continued to be the main source of sugar in Europe, but the Napoleonic Wars between France and Britain led to the blockade of French shipping ports and the rejection of British exports, and as a result the amount of sugar cane that reached continental Europe declined. A German scientist called Andreas Marggraf had discovered in 1747 that large amounts of sugar could also be extracted from a turnip-like plant called *Beta vulgaris* – now better known as sugar beet – that grows in temperate countries. The farming of sugar beet rapidly developed on mainland Europe, and by 1880 beet was the main source of sugar in Europe, with over 300 sugar beet mills operating on the continent. However, by the end of the nineteenth century major developments in

the breeding of new sugar cane varieties, coupled with new farming technology, helped to keep sugar cane in competition with the beet industries, and by 1900 the sugar cane industry had surpassed beet to become once more one of the major economic activities of the entire tropical world. It was not until the First World War, when the supply of sugar cane to Europe was threatened again, that Britain once more began to mass-farm sugar beet, and during the 1920s Britain built 17 factories to produce sugar from beet. Eventually in 1936 the British Sugar Corporation was formed to manage the entire UK sugar beet crop.

Today sugar from both cane and beet is the world's predominant sweetener, with over 130 million tons of sugar consumed worldwide each year. Although they come from quite different plants the sugar crystals from beet and cane look and taste practically identical and the average consumer would not be able to tell them apart. Whatever its origin, the sugar that we consume in our day-to-day lives contributes vital calories to our diet. It is used in cooking, in the preparation of commercially processed foods, as an additive to drinks and as a preservative and fermenting agent. It sweetens without changing the flavour of food and drink. It is cheap to transport, easy to store, and relatively imperishable. In just two centuries sugar has gone from being so expensive that countries would go to war over it, to becoming a simple product found in almost every household. Through this journey it has bolstered the economies of colonial powers and equally devastated indigenous communities. Our global addiction to sugar is still increasing, and in our modern age, sugar cane and beet are two of the most important plants in the economy and nutrition of the world.

Medicine

For as long as plants have been used by humankind for food they have also been used as medicines, to treat pain, aid digestion and promote wellbeing. In fact, evidence suggests that even our primate ancestors were able to self-medicate by eating quantities of berries rich in a chemical called tannin, which has been found to be an effective cure against intestinal parasites. Similarly, gorillas have been found to gather around 118 different plant species with various medicinal properties to promote their health, including the caffeine-rich seeds and fruits of kola trees and the poisonous seeds of the dogbane family Apocynaceae, which in small quantities can help heart conditions. By observing which plants the animals used, and through trial and error, the first hunter-gatherers would have built up knowledge of which plants had curative properties, which would then have been passed from generation to generation. Evidence of this early use in humans has been found in the remains of a 77,000-year-old Stone Age dwelling in KwaZulu-Natal province, South Africa. The rock shelter was found to contain the remains of various different species of rushes and sedges, as well as the leaves of a plant called wild river-quince (*Cryptocarya woodii*), which have been found to contain insecticidal chemicals. The plant material is believed to have been used as bedding material, as the insecticidal leaves would have helped repel mosquitoes.

Over time, increased health and a better quality of life led to greater prosperity for early humans, and as they became more knowledgeable about the plants in their habitat they began to use multiple plants to create complex curative remedies. With increasing complexity came the requirement to document the ingredients and methods of application, and some of these first recorded descriptions of herbal medicine still exist today. Recorded in 1500 BC by the people of ancient Egypt, but believed to have been copied from earlier texts as old as 3000 BC, a few discovered papyrus scrolls provide an insight into the world of Egyptian medicine. The texts describe the use of crude remedies such as a mixture of different herbs on a brick to be inhaled to cure asthma, and the use of half an onion and the froth of beer, impressively described as 'a delightful remedy against death'. Many other ancient cultures also promoted the plant world for its health-giving properties; the Ayurvedic medicine of ancient India describes approximately 1250 different medicinal plants in its Sanskrit texts, and a large number of these have since been scrutinised by contemporary science to substantiate their valuable sources of herbal medicine. From China, the manuscript of *Shennong's Materia Medica*, written during the Han Dynasty, outlines the use of 365 different forms of grass, woods, roots and stones, alongside various animal parts, in their role in health and healing. The ancient Greeks and Romans were also known to use various medicinal plants to promote health.

By the early Middle Ages, the herbal remedies of the ancients were combined with the largely spiritual approach to medicine found in much of medieval Europe at the time. As the church was the centre of literacy during this period, many of the ancient medicinal texts were transcribed by monks alongside their religious manuscripts, and as a result the monasteries soon became the local centres of medicinal knowledge. In their gardens the monks grew many of the necessary herbs and shrubs to create the prescribed cures and potions, such as basil, caraway, garlic and fennel, and people from the surrounding villages would come to the monasteries to receive treatment. One of the major schools of medicine practised by the monks and folk doctors of the Middle Ages was that of humourism, which had its roots in the Greek medicine of 400 BC, and remained the basis of modern medicine in Europe until the eighteenth century. This approach to medicine taught that illness was the result of an imbalance of one of the four humours (yellow bile, phlegm, blood and black bile), and that this imbalance could be set straight with correct administration of the right plants. Plants were believed to have the properties of being hot, cold, moist or dry, and the shape and structure of the plant was believed to be indicative of its healing properties; the skull-shaped seeds of the skullcap plant (*Scutellaria* sp.) were used to treat headaches and the white-spotted lung-like leaves of the lungwort (*Pulmonaria* sp.) were used to treat respiratory ailments.

As a fascination with the healing abilities of plants spread through the scholarly societies of sixteenth-century Europe, institutions were set up by illustrious botanists and doctors with a sole focus the study of medicinal plants. The first such centre was created by the eminent botanist and physician Luca Ghini, who founded the Botanical Gardens of the University of Pisa in 1543, which provided plant material for medical study and the training of doctors and pharmacists in Italy, France and other Western countries. Known as physic gardens, these expansive botanical collections were intended purely for the academic study of medicinal plants, and by 1621 similar collections were founded in Bologna, Cologne, Prague and Oxford. As knowledge was exchanged between these centres of botanical study, the understanding of medical botany expanded, and with the advances of printing technology at the time, comprehensive encyclopaedias of medicinal plants and their uses, called 'herbals', were drawn up. The most famous of the British herbalist authors was John Gerard, whose 1597 herbal, entitled *Generall Historie of Plantes*, described the medicinal uses of both native species and exotic plants from the New World. His text outlined the use of various plants to be ingested directly, to

be used in cooking or to be placed about a person to make use of their healing properties. Some plants were credited with supernatural powers and so considered more effective when collected at a certain phase of the moon or from certain places, and although many of his descriptions are amusing to read in the light of our medical understanding today, many also still

above: Chinese traditional medicine

Ginseng, roseflower, St John's wort, Chinese angelica, cinnamon and dog-rose are just a handful of the plants used in Chinese remedies.

have popular and effective uses. An evergreen shrub called butcher's broom, he said, 'causeth women to have speedie deliverance'; common agrimony was beneficial 'for them that have naughty livers, and for such as pisse blood, upon the diseases of the kidnies'; and *Aloe vera* was beneficial because 'when all purging medicines are hurtfull to the stomacke, Aloes onely is comfortable'.

With the age of exploration and international trade at the dawn of the seventeenth century a wealth of foreign plant species flooded into Europe from Africa, Asia and the Americas. Gradually Europe's botanical gardens began to shift focus from purely beneficial plants. While universities and herbalists still concentrated on the health-promoting nature of the plant world, gardens

above: Malaria cure

The dried bark of the cinchona tree yields many useful medicinal alkaloids, most notably quinine, as discovered by the Quechua of Peru.

—

and organisations began to showcase the most attractive of their new tropical flora, and eventually botany and medicine became two independent schools of study. However, many botanical gardens such as Britain's Royal Botanic Gardens at Kew maintained an ardent focus on both the aesthetic and beneficial qualities of plants, and while their botanical collections displayed the most elaborate flora that was brought back from the tropics, the annals of their herbaria soon began to fill with a multitude of foreign plant species which could be used as food or medicine.

By the early nineteenth century, chemists in Europe had developed techniques to isolate the active chemicals from beneficial plant species, and with this ability scientists and botanists set about scrutinising the powerful contents of a number of known useful plants. At the end of the previous century chemists had become particularly interested in the pain-alleviating effects of the opium poppy, which was driving the opium trade in India at the time, and by 1805 the

German pharmacist Friedrich Sertürner had devised a way to extract the active chemical – which he aptly named morphine, after Morpheus, the god of dreams. Today morphine still remains one of the most effective drugs for the relief of pain. Another important medicinal plant, which came to Europe in the 1600s via the Jesuit missionaries in South America, was the bark of the cinchona tree that had been used for centuries by the Quechua people of Peru to treat shivers from cold. Using it in its unrefined form, the Quechua crushed the bark into a powder which was then diluted in wine. A young Jesuit brother named Agostino Salumbrino who worked in the apothecaries of Lima had the notion that this concoction might also be useful in treating the fevers caused by malaria, and so he hastily sent some samples of cinchona bark back to Rome, where malaria was rife at the time. Amazingly, Brother Salumbrino's hunch proved to be correct, and in 1632 cinchona bark was used to treat the first case of malaria in the Papal city.

As word soon spread of the powerful effects of the bark of the cinchona tree, its use as a prophylactic proliferated throughout numerous malaria-prone areas. As European nations began to establish colonies in Africa, Asia and South America, the demand for the plant increased. However, it was not until pioneering work carried out in 1820 by two French scientists, Pierre Pelletier and Joseph Caventou, that the chemical from the bark which kills the malaria parasite was extracted. This chemical was given the name quinine, derived from the Peruvian name for the bark, which is quina-quina. The revolutionary use of quinine to treat malaria enabled colonialists to push further into previously impassable malaria-ridden regions of tropical Africa and Asia. Quinine was only the first in a number of powerful malaria treatments which would be developed, but as a result of widespread drug resistance to other synthetic drugs, cinchona's active ingredient, quinine, has re-emerged as the medicine of choice to fight some of the most deadly strains of malaria. Today, the only other drug which rivals its effectiveness in treating a wide spread of malarial strains is artemisinin, which is itself derived from another plant called *Artemisia annua*, or sweet wormwood. For many centuries this powerful drug had been known in Chinese medicine under the name qinghaosu, but it was only rediscovered by the Western world in the 1970s.

The barks of particular trees have been found to have other highly important medicinal uses. Originally used by the Egyptians, Greeks and Native Americans for its ability to relieve pain, the willow was the subject of numerous attempts, with varying levels of success, to extract and purify the powerful pain-numbing compounds of its bark

during the 1700s and 1800s. In 1758 an English clergyman named Edward Stone was chewing on a willow twig when its bitter taste struck him as being similar to the bitter tang of the bark of the cinchona tree. Assuming that they might contain similar chemicals, he had some willow twigs dried and powdered and began experimenting with its effects on some parishioners who were suffering from rheumatoid fever. Miraculously, he found that the willow bark had both an anti-inflammatory and an analgesic effect, and his results were later published in the journal *Philosophical Transactions* in 1763. Building on these findings some 60 years later in 1828, a few grams of pure yellow crystals were extracted from willow bark by a professor of pharmacy from the University of Munich called Johann Buchner, and he gave it the name salicin, from the Latin *Salix*, the genus of plants to which willow belongs. By 1829 a French chemist named Henri Leroux had further improved the extraction procedure to produce as much as 25 grams of salicin from 1 kilogram of bark, and soon afterwards an Italian chemist called Raffaele Piria, working in Paris, devised a method of creating an even more powerful substance, which he called salicylic acid. The use of salicylic acid for the treatment of aches and pains and fevers proved to be highly effective, though large doses of it caused irritation to the stomach. German scientists working with the drug decided to add acetyl groups to the acid in order to reduce this painful side effect. Finally, in 1897, a German pharmaceutical company began major trials of the chemical acetylsalicylic acid, or aspirin as it was called, marking the birth of what is now one of the world's most widely used drugs. Today the drug derived from the humble willow is used to treat ailments as far-reaching as strokes, diabetes, cancers, dementia and heart attacks, with an amazing 40,000 tonnes of aspirin consumed globally every year.

Plant Communication

Chapter
Five

'It is only through a better understanding of ecology that we are able to decode many of the interactions of the plant world.'

The plant world provides us with a host of sights and smells. But these bright colours and rich smells are not designed for our human pleasure – they are part of the intricate system of communication that links plants with animals, and plants with their neighbouring plants. For hundreds of years it was believed that plants were silent and inactive in their habitats, but we now know that they have the ability to communicate through soil and air. As an active part of their environments, plants must send out signals to encourage the complex web of interactions that support their ecosystems. Flowers use visual signals and olfactory cues to make their location obvious, as well as to indicate their readiness for pollination. Some use visual signs to warn animals of their toxic chemical compounds to prevent them from being eaten, and some plants may use these signs to give the impression of being toxic. The sugary fruits of plants give off scents to lure animals to disperse their seeds. Plants may also use chemicals to keep away competitive species of other plants nearby, and some use chemicals to warn others of their own species of danger close by.

Our human senses are attuned to what is important to us. Our eyes are able to pick up a certain spectrum of light wavelengths. Our nose and taste buds are honed towards foods that are healthy and nutritious for us and can warn us of toxic or harmful substances. Our ears are fine-tuned to the pitch range of the human voice. In the same way, plants give and receive signals within certain parameters that suit the plant or animal that they are in communication with, and each species is able to signal or respond in its own unique language, be it visual or chemical. Our human receptors can sometimes intercept the conversations of plants where our perceptions overlap the messages of the plant world, and these exchanges are translated to us as the bright colours that we see in flowers, the powerful scent of their blooms, and the distinct aroma of pine forests or the gummy smell of eucalyptus groves. But of course the majority of the communication of the plant world is outside our window of perception, and for that reason a large part of it goes unnoticed by us. It is only through a better understanding of their ecology that we are now able to decode many of the interactions of the plant world.

Scent

Plants create a dominant array of smells that waft through the skies. Summer fills the air with the earthy aroma of tomato plants, the powerful smells of a field of oil-seed, and the pleasant scent of freshly mown grass. Spring bombards our olfactory systems with the heady perfume of lilacs and the morning dew of wet woodlands, and autumn fills the breeze with the distinct smells of damp ground and the mulch of rotting leaves. Rarely, however, do we stop to consider what exactly these smells are telling us about the plants that are releasing them. For example, the potent scent of the night-flowering jasmine (*Cestrum nocturnum*) has been regarded for centuries to have healing properties, and its smell has since been confirmed to increase mental alertness and stimulate the mind.

But of course the sweet scent of jasmine has evolved to promote the survival of this plant species. The unique bouquet of perfume produced by the jasmine, and indeed all other flowers of the plant world, is created to lure one or many specific insect pollinators. In the example of the night-flowering jasmine, its smell has a specific evolutionary function, which is to attract moths

to its flowers. The delicate trumpet-shaped flowers open up at dusk to emit their perfume, which can travel for miles. Moths pick up on this advertisement of the flower's nutritious nectar, and so follow the scent to the flower. Upon arrival the moth gets a meal and in turn the plant gets pollinated. Over time this mutually beneficial relationship would have been reinforced, with specialist adaptations developing in both the moth and the jasmine. Through natural selection their increasingly successful reproduction would have led plants with these flowers to be the dominant variety of jasmine in their habitat – the powerfully scented plants we find today. Likewise the sensory antennae of the moths would have evolved over time to become more receptive to the chemicals of the jasmine scent, allowing them to locate the flowers better than other pollinators. A vast number of plants have invested in producing smells to signal to insects; since antiquity humans have used these scented species to create a wealth of spices,

above: Heavenly perfume

The flowers of the night-flowering jasmine open at night to release their intoxicating scent.

—

above: Hungry beetles

Beetles, such as this Texas flower scarab (*Trichiotinus texanus*) on an Engelmann prickly pear (*Opuntia engelmannii*), seem to favour large, strong-scented flowers.

—

herbs, medicines and perfumes. In humans, the smells that we perceive interact with our olfactory system through the one million or so receptor epithelial cells which line our nose, but insects use their antennae to smell. Where humans sniff, insects wave their antennae to smell the air, and the antennae of some insects are up to 10,000 times more sensitive than the human nose. Honey bees are perhaps the best known pollinators – in the USA alone honey bees have an estimated worth of around $200 billion a year for the work they do in pollinating crops – but the bee is joined by a host of insects that help pollinate the world's flowering plants. Flies and beetles were the original pollinators when flowers first emerged around 140 million years ago, but today butterflies, moths, bees, hoverflies and mosquitoes are all pollinators, to name just a few. The strong scent of flowers enables the insect pollinators to locate them over huge distances, with some species of moths able to detect a flower's scent from over two kilometres away. Protruding in pairs from the heads of all insects, antennae appear in a multitude of different forms, some like feathers, some like paddles and some covered in bristles. Each pair is covered in an array of different-sized sensory hairs, and the most sensitive antenna can bind to a single odour molecule released from a flower in a concentration of just a few parts per million – the equivalent sensitivity of a white-tip shark to single drop of blood. The odour molecules of flowers are released into the air by evaporation, where they are then carried by the wind, creating a scent trail called an odour plume which leads back to the flower. Eventually one of the dispersed molecules will bind with the searching antennae of the right insect. Once it has located the odour plume, an insect such as a moth will fly in a zigzag, crisscrossing through the scent trail to ensure it doesn't lose it, until finally it is close enough to see the flower.

As well as having an incredible ability to pick up smells, many pollinators such as bees are able to distinguish and remember particular odours, and as a response some flowers emit a complex mixture of smells which encourages what is known as flower constancy. This is a behaviour whereby bees choose to forage for food on just one species of flower even when other more nutritious species may be available nearby. Although this is beneficial for the plant, as it ensures that its pollen is effectively transferred to other plants of the same species, it appears to be negative to the insect, as it is encouraged to pass up on potentially more valuable sources of food. Darwin mused on how such a one-sided relationship might have evolved, and in his 1876 paper on the pollination of vegetables he wrote, 'That insects should visit the flowers of the same species as long as they can, is of great importance to the plant ... but no one will suppose that insects act in this manner for the good of the plant. The cause probably lies in insects being thus enabled to work quicker. They have just learnt how to stand in the best position on the flower, and how far and in what direction to insert their proboscides. They act on the same principle as does an artificer who has to make half a dozen engines, and who saves time by making consecutively each wheel and part of them.' Darwin's justification does not however fully explain why pollinators wouldn't learn to stand in the best position on a whole selection of nutritious flowers.

Although flower constancy suggests that insects are being exploited by plants, the insects are still receiving a rich nectar reward for their endeavours. There is, however, a group of plants which use their scent as a way of tricking their insect pollinators into transporting their pollen while not providing a reward. Around one-third of the 30,000 known species of orchids have evolved this ability, and they carry out this trickery largely through their powerful smells, which have developed to resemble the objects of desire for most insects – those of sex and food. One species of tongue orchid called *Cryptostylis subulata* from Australia exudes a smell that exactly mimics the pheromone of a female wasp of the species *Lissopimpla excelsa*. Male wasps of the same species, nicknamed 'dupe wasps', are hoodwinked into thinking that the orchid's flowers are female wasps, which they then hastily mount and copulate with – even to the point of ejaculation – and in doing so they transfer pollen. The male wasp will return to the same species of orchid to repeat the same manoeuvre, and this means he will unknowingly pollinate many hundreds of flowers. Another even more ingenious trick is played by the Chinese orchid *Dendrobium sinense,* endemic to the island of Hainan, which creates a smell that mimics the alarm pheromone of two species of Asian and European honey bee. Predatory hornets of the species *Vespa bicolor* catch this false alarm signal and upon locating the orchid they pounce on it as they would on a wounded bee. Their aggressive pouncing covers their body with the plant's sticky pollen, which they then obligingly carry to the next *D. sinense* orchid that they try and 'attack'.

A single flower can produce upwards of a hundred different chemical odours, which combine to create a complex pattern of smells that change over time to convey a particular message. Day-blooming flowers are able to 'switch off' their scent at night, a mechanism which is assumed to help them save energy, and some night-blooming flowers emit their scent in waves to help it disperse in the night air.

It is not only pleasant smells like roses and honeysuckle that flowers use to attract pollinators; they also create less appetising odours to entice the less salubrious members of the insect world. This has been exploited by a whole group of plants, aptly named carrion flowers, which emit a foul odour like that of rotting flesh as a way of luring scavenging flies and beetles. One carrion-scented flower from Corsica known as the dead-horse arum (*Helicodiceros muscivorus*) emits a stench from its finger-like projection,

opposite: Orchid trickery

The tongue orchid's ability to lure dupe wasps is just one of the amazing ways plants have evolved to maintain their legacy.

—

above: Dead-horse arum

As well as emitting a royal stench these flowers also generate their own heat as a way to attract blowflies.

—

or spathe, which extends from the centre of its arum-shaped flowers that bloom every spring. In doing so it attracts female blowflies looking for a meal or a place to lay their eggs. Botanists speculated for a long time as to what exactly causes this flower's foul aroma, and numerous tests were carried out to find a natural equivalent. In 2002 the results from this study were published in the journal *Nature*, and the offending odour was identified as being from a group of chemicals called oligosulphides, which almost perfectly mimic the stench given off by dead seagulls. Tests were later carried out on the nervous response of the blowfly's antennae to chemicals both from a dead seagull and from the dead-horse arum, discovering that on the basis of their smell alone a blowfly wouldn't be able to tell the difference between the flower and the rotting meat.

Plants need to protect themselves from the plethora of animal herbivores that want to eat them. For plants, protection means powerful chemicals. To deter insects and to ward off herbivorous mammals, plants pack their leaves and green parts full of volatile oils and resins which signal to the animal world that they aren't good to

overleaf: Super senses

Insects locate flowers using their extremely sensitive antennae and finely tuned eyesight.

—

eat. Strangely enough, many of these chemicals are appealing to our human senses – for example the wide variety of flavours found in herbaceous plants, as well as those of garlic, chilli, cloves and mustard. But the aromatic smells of a herb garden and the pleasant aroma of crushed pine are in fact plant defence mechanisms. The volatile chemical resins which produce these powerful signals are secondary metabolites, meaning they are produced secondarily to the basic components required by the plant for growth, and they are collectively known as allelochemicals. The three main compound groups of this chemical arsenal are known as alkaloids, terpenoids and phenolics, and each is utilised by plants in a different way to promote their survival. Some are directly toxic to the animals that eat them, some make the plant indigestible and distasteful, but the most interesting are those compounds which can enlist the help of other animals to protect the plant.

One such plant is the Scots pine (*Pinus sylvestris*), the most widely distributed conifer in the world, which grows as a keystone species in Europe and Asia and provides a backbone on which many other plant and animal species depend. Like all trees, the Scots pine attracts the attention of various types of insects that try and feed on its needles. But the particular scourge of the Scots pine is the European pine sawfly (*Neodiprion sertifer*), which prefers to lay its eggs on the tree's branches, to provide its young with an immediate food supply of pine needles as soon as they hatch. However, as soon as one of these sawflies cuts into the tree's needles in an attempt to insert its eggs, pungent chemicals called terpenes are released from the upper xylem surrounding the wound. Derived from the word turpentine – an aromatic oil produced from tree resin containing these chemicals – there are an estimated 29,000 variations of terpenes produced by different plants, each with a different role in plant communication and defence. The sawfly eggs represent a huge threat to the tree, as the larvae hatching from the eggs can devour large amounts of pine needles, and so the chemical defence of the pine is extremely important. As a first line of protection the flood of terpenes makes the pine needles taste foul to

discourage the larvae of the sawfly from feeding on the pine when they hatch, but while this does seem to partially reduce the damage caused, some larvae are still able to continue feeding.

But the terpenes also provide a second line of defence for the pine. As the terpene chemicals are highly volatile, the moment they are exposed to the air they are carried on the breeze, and soon the pine tree is surrounded with an odorous mist of chemicals. Although invisible to the naked eye the terpene cloud gives off a powerful scent, which even human noses can pick up, producing the distinct aromatic resin smell which is characteristic of pine forests all over the world. But what appears to humans as a quaint fragrance is used by the pine as a powerful message to neighbouring plants and animals. The terpene signal acts first as a trigger for other branches of the tree which is under attack to increase their terpene levels as a way of reducing further egg-laying or feeding attacks from the insects. Then, as the chemical cloud slowly disperses through the forest, it is picked up by a particular species of predatory parasitic wasp (*Chrysonotomyia ruforum*), which is able to use its antennae to locate the original source of the volatile scent. By following the pine's alarm signal the female parasitic wasp flies to the infested tree and quickly sets to work. But the parasitic wasp does not feed on sawfly larvae. Upon arriving at the site where the sawfly have deposited their eggs into the cuts in the pine needles she climbs on top of the eggs and, using her needle-sharp ovipositor, lays her own eggs inside those of the sawfly. Inside the sawfly's eggs the wasp larvae now grow, slowly feeding on the unborn sawflies, and after a couple of weeks the young wasp emerges. So effective are these three-way 'tritrophic' relationships between plants and insects that they can be observed in plant species all over the world. For example, wild mustard releases a volatile called sinigrin when attacked by hungry cabbage-white butterfly larvae, alerting parasitic wasps who come and lay their eggs inside the living caterpillars, resulting in an eruption of live wasps from inside the caterpillar's body. Geraniums, too, produce pungent alkaloids when aphids bite into their leaves and stem, making the plant distasteful while sending a signal to wasps to parasitise the aphids.

The idea that plants can signal to insects has been understood for hundreds of years, but the concept of plants communicating with other plants is fairly new. Work carried out by the German professor Dr Gustav Fecher in 1854

led him to conclude that plants were capable of emotion and could feel pain. Such notions have been repeatedly quashed by the academic community due to a lack of evidence of plants

above: Plant warfare

Plants have evolved numerous and complex chemical defences against prolific herbivores such as aphids.

—

being able to respond to complex signals. However, in the early 1980s scientists began to run experiments to look at the ability of certain trees to 'talk' to one another. One revolutionary three-year field study by US zoologist David Rhoades found that willows which grew close to insect-damaged willows were better protected against insect attack than those growing further away. Similarly, undamaged maple and poplar saplings were found to increase their insect defence chemicals when placed in the proximity of damaged trees. Although these findings seemed to indicate some level of communication between individual plants of the same species, it didn't make evolutionary sense. Throughout the 1980s and 1990s further studies were carried out, and today ecologists believe that while the plant world does 'talk' to the insect world as a way of recruiting predators to remove damaging insects, plants do not communicate directly with their neighbouring plants. What has been found, however, is that neighbouring plants are able to react to signals sent by other plants, thus boosting their own protective chemicals to be better protected against feasting insects. The sagebrush, for instance, emits a volatile defence chemical from its leaves, and when under attack its chemical emissions increase by over 600 per cent. Any tobacco plants growing nearby will pick up on this chemical signal and will consequently boost their own defences.

These airborne signals released by plants and carried by the wind are best suited to travel over large distances and to be received by plants and insects in their larger habitat. But plants can also create short-distance signals to organisms close by, through chemicals sent out into the air and soil in close proximity to another plant likely to be competing for access to its water and nutrients. The ability of plants to limit what can grow near them was first observed in Greece in 300 BC by one of Aristotle's pupils called Theophrastus, who noticed that chickpea plants 'exhausted' the soil around where they grew, destroying other plants. Pliny the Elder also remarked on this behaviour in the first century AD, observing that barley and walnut plants appeared to 'scorch up' the soil, preventing the growth of other plants close by. In 1937 an Austrian professor named Hans Molisch coined the term allelopathy to describe the ability of plants to influence those around them, from the Greek words *allelon* 'of each other' and *pathos* 'suffering', and since then a huge body of work has

'Through the process of natural selection, the forms of flowers diversified.'

above: Honeysuckle

Lonicera periclymenum, or common honeysuckle, is native to Europe's hedgerows, scrublands and woodlands.

—

looked into how plants are able to use chemicals to inhibit others. It is now known that allelopathic chemicals are released as a sort of natural herbicide by the administering plant, with the effect of limiting the shoot and root growth of competing plant species. Depending on the species, these chemicals can be present in the leaves, flowers, roots, fruits, seeds or stem of the plant, as well as in the soil surrounding it.

Perhaps one of the most well-known groups of allelopathic plants are the eucalyptuses or gum trees, which grow natively in Australia but have since been planted in gardens and parks all around the world. The distinctive blue-grey leaves and bark of eucalyptus trees are full of volatile oils and allelochemicals, which give these trees their characteristic menthol aroma and make them highly poisonous to most animals. These trees produce their oil in such quantities that in the right conditions it evaporates and fills the atmosphere with a blue cloud of finely dispersed oil droplets. In 1788 the name of Australia's eucalyptus-covered Carmarthen and Lansdowne Hills was appropriately changed to the Blue Mountains. When the oil-filled leaves of the eucalyptus fall to the ground they begin to rot, and with the help of rainwater their allelochemicals leach out into the soil, where they are reinforced by the further release of chemicals from the tree's roots. Growth of any potentially competitive plants will be suppressed in the soil around the tree.

Colour

Together with their scent, the size, shape and colour of different flowers all help to attract a huge range of pollinators, including bees, flies, moths, butterflies and beetles, as well as a whole manner of small mammals and birds. But flying insects are the primary pollinators of the world's flowering plants, feeding on their nectar, transferring pollen as they go, which aids the reproduction of around 70 per cent of the world's flowering plants. However, insects have no concern for whether they transfer pollen to the right flowers or not, so the flower needs to ensure that it makes its sugary food more obvious and appealing than the next species of plant in order to increase the likelihood of a return visit. The bigger the flower's advertisement the more attention it is likely to receive from pollinators, but evolution towards infinitely larger flowers would force flowering plants to reduce their fitness in other areas of their growth, such as the size of their roots or the number of seeds they can create. If a plant invested most of its energy in simply producing the biggest possible flowers, it might have little energy left for anti-herbivore defences such as filling its petals with distasteful chemicals or shielding them with spines. Flowers therefore have evolved to make a compromise, between producing the 'ideal' flower which will attract the most pollinators, and one that is economical to produce while allowing enough resources to be invested in the plant's fitness as a whole. This is the basic principle that gives rise to every shape and size that we see around us in the natural world today.

But as all plants in any one habitat are constrained by the same limiting factors of the amount of light, nutrients and water that they can absorb, why has this economic model not led all plants that live together to produce identical flowers of the same size, shape and colour? The natural world

opposite: Hummingbird
pollinators

Different species of flower dust birds
at different parts of their bodies so
that the pollen will be more likely to
end up on another plant of
the same species.

—

is a system based on competition. In this hypothetical landscape of uniform flowers, insects would indiscriminately fly from plant to plant and there would be a high chance that they wouldn't cross-pollinate two of the same species, as no flower would appeal to them more than the next. For the very first flowering plants that appeared on Earth over 140 million years this wouldn't have been such a problem, as their novelty alone as the only flowers in the ancient landscape would have been enough to attract pollinators. But natural selection drove the plant world to eventually diversify into a whole array of novel flowers to grab the attention of particular pollinators.

Plants, through the process of natural selection, began to unknowingly experiment with the forms of their flowers. The reshuffling of a plant's genes with the start of each new generation gave rise to occasional mutations in the size of its flowers, the layout of its petals or the colour of its blooms, and some of these novel changes grabbed the attention of particular animals. For a large part it was the brighter and more obvious flowers which would have been located by insects more regularly than flowers with inconspicuous petals, and so their pollen would have been transferred more regularly. In the cases where the showiest flowers also offered a nutritious, sugary reward to the insects that visited them, their attraction to them would have been reinforced. Over time, more pollinators visiting a certain type of flower results in more chances of cross-pollination, fertilisation and the production of more seeds, and the populations of these plants would have increased in numbers. The animal kingdom shows preferences towards particular shapes and colours when they come to selecting which flowers they will visit, and flowering plants have evolved to exploit what works best for their own particular pollinators. Their flowers have been shaped accordingly, causing the immense diversity of all flowering plants today.

Hummingbirds are an important pollinator in the tropical habitats where they live, drinking nectar from large recurved, trumpet-shaped flowers which can accommodate their long thin beaks. They have good colour vision which extends into the near-ultraviolet end of the spectrum, but they have a relatively poor sense

(a) *Rudbeckia*
visible light

The human eye sees this flower
as having all yellow petals.

(b) *Rudbeckia*
UV light

In UV light reveals
that the petals reflect two
separate colours.

(c) Foxglove
visible light

In visible light we only notice the
white and purple markings of
the petals.

(d) Foxglove
UV light

In UV light it is possible to see the
defined 'landing strips' leading into
the flower's throat.

of smell, and as a result many of their attractant flowers tend to be relatively odourless. The species they pollinate instead appeal to their eyesight with red, orange and bright pink petals, such as the vivid flowers of the coral tree and agaves of South America. Many flying insects have poor vision at the red end of the spectrum and so pass up on these red flowers, creating a convenient partnership between the hummingbird and the flower. Some species, such as the scarlet gilia (*Ipomopsis aggregata*) from Arizona, can actually change the colour of their petals over the course of a season to exploit the occurrence of different pollinators at different times of the year. In mid-July the flowers are a dark red in order to attract hummingbirds, but as the birds begin to emigrate in August, the flowers shift from red to pink to white, to attract the attention of the pollinating hawkmoths (*Hyles lineata*), which remain in the area when the hummingbirds have left.

Moths are highly important pollinators, attracted to various species of plants that open their flowers at night. These include many of the night-flowering cacti, which often produce a strong odour to help the moths locate them in the dark. Many of these species also have large white or cream petals which reflect the moonlight, to help them stand out in the desert landscapes where they grow. Similarly, some night-blooming flowers that grow in the tropical forest of the Philippines, such as the jade vine (*Strongylodon macrobotrys*), have ghostly green flowers which appear to glow in the moonlight, which is assumed to make them more visible to their bat pollinators. Beetles have relatively poor colour vision and so are useful pollinators of less vivid, greenish or off-white flowers which lure the beetles primarily using strong scent. Conversely, butterflies tend to have a weak sense of smell but are attracted to large red and orange flowers.

Our human eyes are only sensitive to a very small portion of the whole electromagnetic spectrum, allowing us to see the colourful wavelengths between about 390 and 750 nanometres in length ('the visible spectrum'), while the other trillion wavelengths remain invisible to us. As a result we can only absorb the visual messages of colour and pattern that plants produce within this window. However, we know that the most important group of flower pollinators,

(a)

(b)

(c)

(d)

the bees, have a very different colour detection system and as a result are able to see both visible and ultraviolet (UV) wavelengths of light. Unseen to our eyes, some flowers use the reflectance of these UV wavelengths to create a whole other level of floral communication, which until fairly recently remained hidden to humans.

The initial work which discovered the ability of insects to see in the UV wavelengths came from studies carried out on ants over a century ago. Just a few decades later, in 1924, the experiments of one German scientist, Alfred Kühn, attempted to train bees to recognise flowers in different types of light, and revealed that they too have UV vision. Since then a great deal of work has been carried out to explain how plants exploit the UV perception of bees to attract them to their flowers, and with the advancements of camera technology over the last 50 years scientists and photographers are now able to create photographs which reveal these hidden UV patterns. These images are the closest that we can get to visualising how bees see the world, and they reveal the multitude of previously unseen mechanisms that flowers use to attract these types of pollinators. In total about 7 per cent of flowers have been found to posses patterns only seen by animals with UV-receptive eyes, and a large proportion of these are plants with large blooms. Some striking examples include *Geranium clarkei*, which appears to our human eyes as a faint pink flower with pink striations pointing inwards towards its centre. However, when its image is overlaid to include the reflectance of UV wavelengths, the flower is revealed as a bee would see it: a dark purple head with black stripes almost like a landing strip leading the pollinator in towards the pollen in the centre which surrounds its bright glowing nectaries. These landing-strip patterns guide bees to the plant's nectar, and they are even more pronounced in the flowers of the foxglove (*Digitalis purpurea*): in visible light these plants have a large white trumpet-shaped bloom with purple markings like splattered paint leading into the throat of the flower, suggesting a mechanism to attract types of non-UV-sighted pollinators as well, but in UV light the flower is transformed. The petals glow a vivid purple colour and the splattered markings all but disappear. In their place, two bright stripes of white lead deep into the flower's throat. Another widely seen UV pattern for drawing bees to flowers is perhaps best exemplified in the flowers of *Rudbeckia fulgida* or orange coneflower from

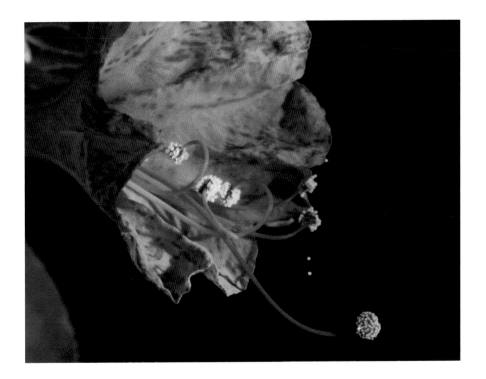

North America. To the human eye this plant has star-shaped flowers with vibrant yellow petals which act like insect landing pads on the end of stiff metre-long stalks, but in the

above: Glowing pollen

The pollen of *Mirabilis jalapa* emits a strong fluorescent glow, which is thought to attract nocturnal moths.

—

pattern of its UV reflectance the yellow petals reveal an elegant trick. What appears on the petals in visible light to be all one colour, made up of the same yellow pigment, under UV light reveals itself to be made of two very different cells, with those close to the centre of the flower absorbing UV and those at the tips of the petals reflecting UV. The UV absorption is caused by chemical pigments in the petals called flavonoids and as a result the flower as a whole appears two-toned, with its outer parts retaining their yellow colour while the inner parts manifest themselves as matte black. This gives the flower a large black bull's-eye target, which is thought to make it easier to locate and feed from.

Behavioural studies in the twenty-first century have sought to provide a better understanding of the mechanisms by which bees forage. We know now that they do not 'see' shapes or objects but instead detect parameters and recognise places, and by using their 300-degree vision they are able to triangulate on just a few clues in order to find food. The patterns that we see in the flowers around us have evolved to play to such perception, and as our understanding of both plant and pollinator increases we are able to gradually unfold more details of the complex relationships that have formed between them. The yellow and black of the *Rudbeckia* petals is a useful clue to help us understand how bees respond to the colour signals from plants, as it tells us that the contrast between colours plays a significant role. There appears to be yet more evidence for the importance of this colour contrast, in the way that non-floral parts of the plant are seen, or not seen, by bees. As the green parts of a plant must be able to absorb light from the sun in order to photosynthesise, much of the UV light that falls on the leaves and stem is absorbed by pigments such as flavanoids and chlorophyll. As a result, for an animal who sees predominantly in the UV region of the spectrum, green vegetation appears almost black. The effect of this is that the UV-reflecting parts of flowers are heightened by the black background, making them more obvious to certain pollinators. But the UV vision of insects is not always exploited in a mutually beneficial way by plants, and some carnivorous plants have been suggested to lure insects with their UV glow. One such species is the sundew (*Drosera longifolia*), which produces droplets of clear sticky mucilage from its green leaves. For insects with UV vision its leaves appear bright red, edged with globules of shining liquid. When insects come to land on the appealing liquid they are caught and slowly digested by the plant.

In some species such as the sunflower and horse chestnut, UV patterns can change within the flower's life span. For the insects that read these changes a bright UV pattern can indicate a flower that is full of nectar while a faded UV pattern can mean a flower has already been pollinated and no longer contains food. Another occasion when the timing of the message which a plant conveys is particularly crucial is when it comes to attracting seed dispersers to its fruit. While they may be a delicious foodstuff for humans, the primary importance of fruit is as a reliable method of dispersing a plant's mature seeds via a host of hungry frugivores. As successful seed dispersal is vitally important for the survival of a plant species, fruits have been exposed to a high level of selective pressure over their evolutionary history, and as a result they exist in a dazzling array of shapes, tastes and appearances. In the same way that a flower attracts its pollinators by offering a sugary reward, a plant will attract seed dispersers by encasing its seeds in a nutritious bundle. As this sugary flesh is eaten the seeds will be spread throughout the habitat to eventually germinate in a new location. Often before a fruit is ripe it is distasteful and inconspicuous, such as the fruits of a raspberry plant which begin their lives green and bitter-tasting. At this stage the seeds within the fruit would not be developed enough to germinate and so being eaten by fruit-loving birds at this time would be detrimental to their survival. But as the raspberry's seeds approach maturity the fleshy body surrounding them begins to fill with sugars, and this in turn causes colourful pigments called anthocyanins to flood into the fruit. As the fruit ripens its colour changes, and over millions of years the animals which have evolved along-side it have adapted to register this change of colour as a signal that the fruit is now nutritious and good to be eaten.

above: Bright berries

The vivid red berries of *Anthurium plowmanii* from South America are eaten by birds that spread them in their droppings.

For the large part the fruits we see in the plant kingdom today are the most nutritious and beneficial that have been favoured by seed-dispersing animals, as any new form of fruit that emerged throughout the evolution of seed plants that was unappetising simply wouldn't have been eaten as readily, and as a result its seeds would have dispersed less successfully. The colours of fruits and their structures have evolved to communicate the quality of the reward within as a way of securing the attention of the most discerning palate. An example of this concerns the redwing, a fruit-eating bird which has worked out that the darker variations of the wild fruits from its habitat in Iceland contains the highest levels of antioxidants – and so they preferentially eat these over any other. Many of these examples highlight beneficial exchanges between plants and animals that help both giver and receiver. While this is a great strategy for a plant to ensure cooperation from its animal helpers, plants could save resources if they were able to trick a pollinator or seed disperser into helping them without producing rewards. Whether such deception could be pulled off by a plant has been a matter of contention among scientists for over a century. Charles Darwin found it hard to believe, noting in his 1862 book on orchid fertilisation that anyone who believed in 'such an organised system

above: Insect infestation?

The black flecks on the leaves and stems of this species of *Paspalum* grass are thought to mimic ants and thus deter herbivores.

—

of deception ... must rank the sense or instinctive knowledge of many kinds of insects, even bees, very low in the scale.' But even Darwin could provide no other explanation for the handful of British orchids which appeared to repeatedly lure pollinators without providing any nectar. For 23 days he examined flowers for any sign of nectar: after rain, after exposure to sun, at night, and even after dissection – and still not a single drop of nectar could be found. Since Darwin's day many hundreds of studies have been conducted on plants across the globe to explore the extent to which plants are able to trick their pollinators and dispersers, as we have seen earlier in this chapter.

But the world of plants has evolved a whole array of other hoaxes. Some plants have evolved tricks which will get animals to disperse un-nutritious seeds. This is believed to be the case with the saga tree (*Adenanthera pavonina*) from India's tropical rainforests, whose spiral seed pods split open as they ripen to reveal their bright red seeds. These berries are actually fairly poor in nutrients, and are merely mimicking other fleshy seeds. In this way they can benefit from a seed dispersal service without investing as much energy in producing nourishing seeds.

As well as tricking animals into visiting their flowers and fruits, plants can also trick animals into staying away. Some plants make themselves look like a particular poisonous species while others can cause their leaves to appear ill and damaged. These masquerades act to reduce the likelihood of the plant being eaten by herbivores and therefore they increase the plant's chance of survival. But for a plant to successfully convey false information about itself to any animal, the animal must have already formed an association with the plant or organism that is being portrayed, and only then can another species evolve to exploit that association. In this way harmless hoverflies have evolved to mimic harmful wasps by having yellow and black striped bodies. For small mammals that forage on the fleshy stems of small shrubs an infestation of aphids or ants is enough to make them avoid a food plant, and as a result there is a handful of plants that have evolved patterns of dark flecks and dots that appear, even under close scrutiny, to look like an insect infestation. Perhaps one of the most convincing examples is seen in a species of European grass called *Paspalum paspaloides*, which has small black floral structures a few millimetres long dangling from its flowers to trick even the sharpest eye into thinking it is covered in black aphids. Other plants masquerade as having sustained damaged from an infestation. The white and green patterns of leaf variegation, as seen in many hundreds of plant species, has long been thought to have a role in making a plant look less appetising by resembling insect damage, but some exaggerated examples have been noted in which patterns appear to imitate damage from a specific insect. Scientists working in southern Ecuador found that some leaves of *Caladium steudneriifolium* exhibited white patterns resembling the trails of the mining moth caterpillar, and as a result these leaves were eaten by herbivores far less than green leaves of the same species which grew nearby.

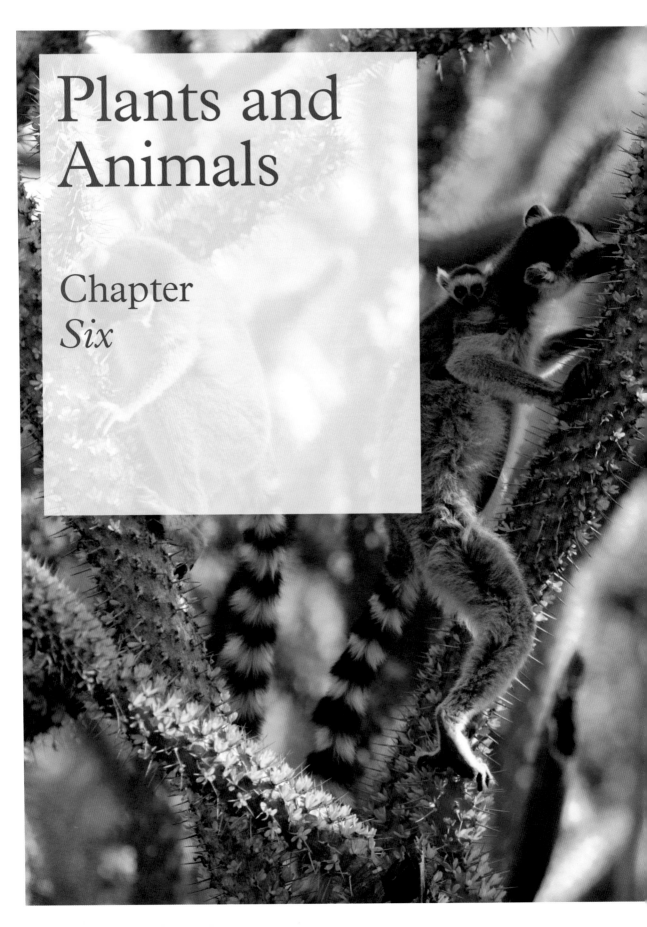

Plants and Animals

Chapter
Six

'The proliferation and evolution of animals increases the selective pressure on plant life.'

Throughout millions of years of evolution animals have developed important relationships with plants, resulting in an interdependent web of floral and faunal species – together with innumerable species of fungi and micro-organisms – that have come to rely on one another for survival. Animals in all habitats of Earth's biosphere have found ways of using plants for food and shelter, and in turn plants have evolved to use an array of animals to pollinate and disperse their seeds, as well as for protection and nutrients. These co-evolved relationships between plant and animal define the broader communities of species that can survive in any given environment, and have helped shape the evolution of life on our planet, affecting which species adapt and change and which species become extinct. Most importantly, the study of relationships between plants and animals tells us that biodiversity – the totality of different organisms of both plant and animal, the genes they contain, and the communities they form – must be maintained if we are to retain a healthy and stable environment on our planet.

The importance of innumerable plant and animal interactions which define the world's wild habitats has been well documented since the birth of the early environmentalists in the nineteenth century. One such environmentalist was the Scottish-born American naturalist John Muir, who was one of the first advocates for the preservation of wilderness areas in the USA. Described by author William Anderson as exemplifying 'the archetype of our oneness with the Earth', Muir is also credited as being the first to highlight the interwoven nature of all plants and animals in the natural environment. In 1911 he famously wrote of the natural world, 'When we try to pick anything out by itself, we find it hitched to everything else in the Universe.' But long before Muir captured America's attention with his musings on the importance of man's place in nature, the fundamental importance of maintaining the balance between plants and animals had been understood by indigenous cultures all over the globe. Their spiritual approach to nature is still practised in the native lands of these people today – in the deserts of sub-Saharan Africa, in the rainforests of Asia and in the jungles of South America. These communities have lived in balance with the plants and animals around them for many hundreds of years by upholding the ecosystem services and natural resources on which they rely. Many of these indigenous groups have been found to manipulate their environment to boost its heterogeneity, such as the Numic-speaking people of California's Great Basin. They are taught to whip the trees they gather pine nuts from, breaking off the tips of the branches, which in time sprout into new trees and support more biodiversity. The activities of some indigenous communities have even been found to restore biodiversity in degraded landscapes, such as in Oaxaca, Mexico, where low-level farming of the local forest areas helps renew the landscape and encourages more plants and animals to colonise the environment.

The relationships that have evolved between plants and animals to assemble our biodiverse habitats take shape in a multitude of forms. A plant or animal may indiscriminately use various different species for the same purpose – ants take shelter in the folds of a range of different leaves, primates may swing from the branches of a whole array of trees to escape the predators of the forest floor, and birds may use twigs and branches from a plethora of different plants to create their homes. Any one of the plants and animals that benefit from this type of relationship could be out-competed by another, and it

would simply be replaced by another similar one. As well as these loose affiliations, there are also a great number of partnerships which exist whereby one or more of the plants and animals involved requires the other species for its survival. These types of one-sided relationships are described as commensalistic. One example is the way that some plants disperse their seeds by hitchhiking a ride on the fur or feathers of animals, as practised by the Velcro-like seeds of the cocklebur (*Xanthium*) – its hook-like structures attach onto a bird or mammal which disperses it away from the parent plant; the animal is neither harmed nor benefited. However, the majority of relationships that exist between the plant and animal worlds have acquired a higher level of formality, wherein both parties benefit from each other. These mutualistic enterprises have evolved over millions of years.

Pollination is one of the most obvious forms of the mutualistic relationships that develop between plants and animals, as explored in Chapter Five, and over time this mutual interest causes body shapes to naturally evolve,

above: Hitchhikers

Many seeds, such as those of *Pisonia grandis*, attach themselves to birds' feathers to aid their dispersal.

—

above: Pollen transport

As well as being covered in hairs, bees are also electrostatically charged, helping pollen stick to them.

—

as shown by moths with long proboscises pollinating orchids with long nectaries, and bright showy flowers attracting visually searching bees. The adaptations which have evolved in common bumblebees (*Bombus* sp.) have made them some of the planet's super-pollinators, able to carry pollen equal to 90 per cent of their own weight, tirelessly visiting flowers from sunup to sundown. Such a pollen-carrying capacity requires the bumblebee to be of a particular girth, and in response many flowers have evolved to be particularly sturdy to accommodate the bumblebees while they feed. On the island of Madagascar, the handsome black-and-white ruffed lemur (*Varecia variegata*) is the primary pollinator of a species of tree called the traveller's palm (*Ravenala madagascariensis*), which is in fact a member of the bird-of-paradise family Strelitziaceae. Its confusing name is a result of its enormous palm-like, paddle-shaped leaves which store rainwater in the chamber made by the base of the leaf petiole. In Madagascar's lowland and mid-altitude rainforests which run along the east of the island lemurs clamber up the 15-metre-high trunks of the traveller's palm to reach its flowering parts. Using their dexterous hands they open up the flowers, allowing them to get their pointed muzzles and long tongues in to reach the sugary nectar within. In doing so their fluffy faces are covered with pollen, which is subsequently carried from flower to flower as they feed. The flowers are cross-pollinated, and in turn the lemurs ensure that future generations of the traveller's palm will exist, which will help feed their own progeny. Weighing between 3 and 4.5 kilograms, the black-and-white ruffed lemur is the largest pollinator in the world – no other animal on Madagascar has the strength or dexterity to pollinate the traveller's palm. Due to deforestation for timber and agriculture, however, these lemurs are losing not only their homes but also their source of food, and as a result the survival of both tree and lemur hangs in the balance.

above: *Banksia*

The cone-shaped flower heads of these plants provide food for a whole host of animals.

—

In the coastal heath sand plains of southwestern Australia another small mammal called the honey possum (*Tarsipes rostratus*) has an important relationship with the plants it pollinates. The honey possum's name is also misleading, as – unrelated to the possum family – it is the last remaining member of an ancient group of marsupials called Tarsipedidae. The noolbenger – its native Australian name – is a mouse-like marsupial with a pointed snout and long tongue. It is one of the few mammals known to survive entirely on a diet of nectar. Although its interactions with the plant world have been the focus of a great deal of research since the 1970s, it is still not fully understood. It is primarily the flowers of the genus *Banksia* which are thought to provide this creature's sugary diet. Named after Sir Joseph Banks, the first unofficial director of the Royal Botanic Gardens at Kew, the first *Banksia* plants were discovered during Captain Cook's maiden voyage in 1770. The inflorescences of these plants are held in dense cylindrical spikes resembling the flower head of an artichoke, creating a bloom filled with nectar. The noolbenger is able to delve into the nectaries of each flower, extracting the sugar with a specially modified tongue with a bristled tip. These marsupials weigh in at little more than 10 grams, and the nectar they gather from these plant species provides a plentiful food source. As long as at least one of their favoured plants is in flower in their

previous: Devil's claw

The seed capsules of this species are
adapted to hooking onto the legs of
passing animals.

—

habitat at any given time, their population numbers will remain stable. However, their tiny body size means that even a brief period of decreased food supply can cause a population to plummet to near extinction, as they are unable to switch to alternative food sources such as fruits or insects.

The marsupial carries the plant's pollen stuck to its snout, whiskers and fur. Banksias are also, however, visited by a selection of other pollinators, including small nectar-feeding birds such as purple-crowned lorikeets, red wattlebirds and New Holland honeyeaters, as well as various insects and other small mammals. So while the noolbenger could not survive without nectar-producing *Banksia*, it is not known whether the noolbenger's role as a pollinator is essential to the plant.

In habitats where the populations of animals may fluctuate unexpectedly, the primary pollinator of a plant may switch over time. A plant that has evolved alongside a specific species of bird or bee for hundreds of thousands of years may suddenly lose its pollinator due to habitat loss or competition from other animals, and may soon find that anything from a mammal to a moth has moved in as its pollinator. Lizards rarely visit flowers for food in temperate climates, and yet a bizarre set of events on the island of Mauritius in the Indian Ocean has caused one species of gecko to take up the surprising role of key pollinator for a tropical tree. The tree is the threatened *Trochetia blackburniana*, which has historically been pollinated by a nectar-drinking bird called the olive white-eye (*Zosterops chloronothos*). However, due to predation from animals introduced to the island, its numbers have plummeted from an estimated 350 pairs in the 1970s to potentially as few as 190 pairs today, in a habitat measuring only just over 60 square kilometres. Researchers from the University of Zurich in Switzerland studying the endangered tree observed that various insects also visited its red flowers to drink its nectar, giving them hope that perhaps the insects could provide a pollination service where the birds were no longer able to. On closer inspection, however, these bugs weren't carrying any significant amounts of pollen, ruling them out as a possible lifeline for the tree. They also noted that a brilliant green and blue species of day gecko would occasionally visit the tree's flowers to drink the nectar, and in doing so the gecko's chest,

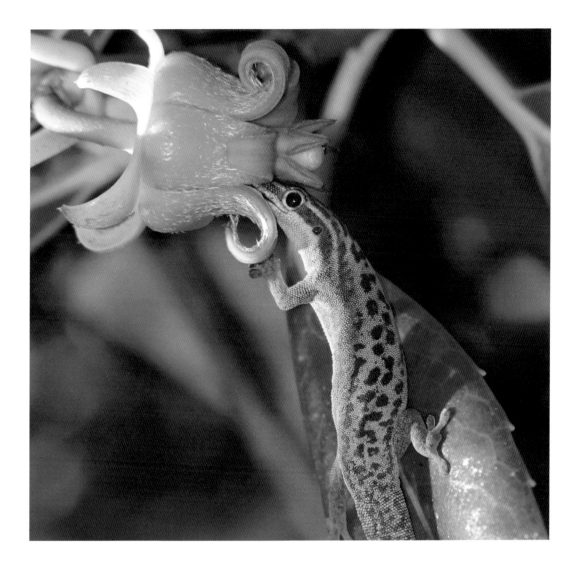

throat and head would become covered in pollen. However, it appeared that climbing out onto the tree's branches exposed the gecko to predation by the Mauritius kestrel (*Falco punctatus*), and so, not wanting to be eaten, the geckos were deterred from returning to the flowers. Miraculously, however, groups of *T. blackburniana* trees were found growing among thickets of dense palm-like screwpine (*Pandanus* spp.), which provided a microhabitat for high numbers of geckos. The spiky lattice of fronds provided cover from the preying eyes of kestrels, allowing the geckos to freely make regular visits to the tree. The researchers

above: Lizard visitor

The blue-tailed day gecko (*Phelsuma cepediana*) transfers both the pollen and seeds of the critically endangered *Roussea simplex*.

—

found that the coexistence of these two plants gave the gecko the perfect habitat to drink the flower's nectar and spread its pollen from tree to tree, ultimately helping save *T. blackburniana* from the brink of extinction.

Reptiles have forged important relationships with a number of plant species around the world, with 95 per cent of these occurring in island habitats. In many of these isolated environments where bird and insect pollinators are in decline due to habitat loss or other population pressures, it is believed that flower-visiting lizards may be the key to ensuring the survival of a number of plant species. On the islands of the Fernando de Noronha Archipelago off the coast of Brazil a 10-centimetre-long dark-spotted lizard called the Noronha skink (*Trachylepis atlantica*) drinks the nectar and pollinates the flowers of the mulungu tree (*Erythrina velutina*). The bark and leaves of this tree have been used as a powerful sedative for hundreds of years and today it is one of the most popular natural tranquilisers worldwide. As the tree flowers during the islands' dry season the moisture trapped in its flowers also provides the skink with a vital source of fresh water. In New Zealand the *Hoplodactylus* geckos are important pollinators of many of the island's native plants, and the New Zealand Christmas tree flower (*Metrosideros excelsa*) from the North Island is also known to be pollinated by as many as 50 different species of gecko. The role of lizards in pollination has been underestimated for many years, as has their role as important agents of seed dispersal, primarily because they indulge in a largely carnivorous diet. However, the ability of lizards to reach high densities on islands, with relatively low predation rates compared to the mainland, means that they are able to modify their diet to include fruit as well as nectar and pollen.

A plant must spread its seeds away from its body to survive, and it achieves this through a whole host of different mechanisms – producing seeds with wings to float on the breeze, making seeds with buoyant bodies to drift along watercourses, and forming seeds with structures which lead them to be moved from one place to another by animals. The Swiss inventor George de Mestral had a revelatory moment upon finding his dog covered in the hooky burs of the burdock plant (*Arctium* sp.), inspiring his invention of Velcro. The sticky nature of some of these seeds provides plants with an efficient system

opposite: Organ-pipe cactus

This mighty desert species plays an important role in its ecosystem, where it is pollinated by bats.

—

of dispersal, and eventually the unwitting carrier will brush or rub off the seeds where they will hopefully be able to germinate.

Most plants require the help of separate animal species for the dispersal of their seeds and the pollination of their flowers. For example, the Brazil nut tree (*Bertholletia excelsa*) growing in the Amazon rainforest is reliant on ground-dwelling agoutis for the dispersal of its seeds, as these are the only animals with teeth strong enough to crack open the tree's seed pods, and for its pollination it relies on large euglossine bees. However, for a handful of species of columnar cacti the animals that pollinate their flowers are the same as those that disperse their seeds. One of these species is the impressive organ-pipe cactus (*Stenocereus thurberi*) which grows in the deserts of the American southwest. Its survival is intrinsically linked to the migratory lesser long-nosed bat (*Leptonycteris curasoae*). Growing to heights of 8 metres, the organ-pipe cactus is one of the largest species of cactus on the planet. In its habitat, which ranges from southwestern Arizona to the western Sonoran desert in Mexico, these slow-growing giants can live for up to 150 years, forming dense clumps of curving vertical stems resembling the pipes of an organ. Every spring organ-pipe cacti older than 35 years sprout small bud-like growths at the top of their tall stems, developing into tough 8-centimetre-long cone-like flowers. To prevent them from drying out in the intense heat, the flowers only open after the sun has set, when the conical buds peel open to reveal creamy petals surrounding pollen-laden anthers. As the eyes of the cactus's bat pollinators contain predominantly rod photoreceptor cells they have almost no ability to see colours, and as a result the cactus flowers have no need for lavish displays of colour. Various species of bat-pollinated flowers therefore tend to be white, green or sometimes purplish-red or pink. Each organ-pipe cactus develops upwards of 100 flower buds every season, with as many as 20 flowers opening per plant in one night, which gives a large region of these plants a total flowering period of many weeks between May and July. The bats will often use the light of the moon reflected off the white petals to locate the flowers at close range, but it is their well-developed sense of smell which brings them to the cactus from afar – the organ-pipe cactus flower emits a particularly pungent musky or nutty odour.

The lesser long-nosed bat does not live around the same organ-pipe cacti all year round, as it is one of the few migratory species of bats, making an annual trip in excess of 1800 km from their winter habitat to their summer roosting grounds. In North America this species of bat divides itself into three separate groups, each embarking on their own routes of migration, with two of them giving birth to their pups in the springtime in either coastal or montane habitats and the other giving birth in the winter. But for all three groups of bats to successfully complete their three-month journey across the expanse of the desert they must time their travel perfectly to coincide with the annual flowering and fruiting of the organ-pipe cactus. Every year hundreds of thousands of lesser long-nosed bats leave their warm roosting caves just after dusk and set off in search of cacti along their migratory corridor. After they have located a stand of flowering organ-pipe cacti they hover over the flowers and dip their pointed heads into the nectar-filled cones to feed, using their long brush-tipped tongues and being dusted in pollen in the

above: Security guards

The nectar produced by this species of *Opuntia* attracts ants which help keep insect herbivores at bay.

—

process. For five hours every night they can feed on flowers over a distance of around 100 kilometres, transferring pollen over this massive distance as they go. The metabolism of these bats is so high that in order to travel the great distances required for their migration they must drink nectar from between 80 and 100 flowers per night, making them a highly efficient pollinator for the cacti they feed from. By the time the sun comes up, and with it the desert heat, the flowers of the organ-pipe cactus have closed up to prevent them from drying out and the bats have settled down in their temporary roosts. This cycle of flowering and feeding is repeated every night for many weeks, and by the end of its migration a single bat will have helped pollinate many thousands of flowers in return for the sugary food which will nourish the growing pups of the pregnant females.

Following a gestation period of six months, which overlaps with their migration, the females give birth to a single pup, which will be weaned in their maternity roosts at the end of their migration on milk enriched by the sugars of the organ-pipe cactus. While the bats are tending to their newborns the ovaries of the successfully pollinated cactus flowers begin to swell and red-fleshy fruits covered in spines emerge. As the cricket-ball-sized fruits ripen they lose their

protective spikes and coinciding with the late summer rains they begin to split open, revealing their red seed-filled pulp. By late summer/early autumn the resources of the bats' maternity roosts have diminished, and in order to survive the young pups and their parents set off on the return leg of their annual journey. This time it is the sweet fruit of the organ-pipe cactus which fuels their voyage. As they feed on the fruits along their migratory route the bats deposit seeds over a vast range. These will germinate and, in time, with the help of nurse plants or protective rocks, will grow into the next generation of organ-pipe cacti.

Though the extreme temperatures and barren landscapes of deserts give the impression of a lifeless environment, they are actually areas of high biodiversity. In the arid desert environment where water is scarce, herbivores such as the pig-like javelina, desert tortoise and various insects seek out the fleshy stems of plants as a vital source of water. One of the most efficient water-storing cacti is the fish-hook barrel cactus (*Ferocactus wislizeni*). Even though its spines are sufficient to prevent it from being feasted on by most herbivores, it is still scourged by the caterpillars of tobacco hornworm (*Manduca sexta*). As a second line of defence against the caterpillar the barrel cactus has formed an alliance with a certain species of desert ant from the genus *Crematogaster*, more commonly known as acrobat ants. The cactus oozes a sugary fluid from small glands nestled in between its spines called extrafloral nectaries, which attracts the acrobat ants. Happily taking up refuge on the outside of the cactus, the ants feast on the sugar and in turn help protect the cactus by keeping away any preying hornworm larvae.

But where some plants have evolved partnerships with insects to help protect them, other plants have developed an alternative strategy for survival. Carnivorous plants have developed various trapping mechanisms to lure and ensnare insects. They grow in areas with poor soil quality, in a vast number of habitats ranging from the moist swamps of North and South Carolina to Borneo, Australia and Siberia. In each of these different landscapes these plants get their necessary diet of nitrogen by consuming the bodies of insects, frogs and in some cases small mammals. Each species has evolved its own tactics of entrapment, enticing their prey using colours, scents and the promise of a meal.

These mechanisms have enthralled botanists and plant collectors for centuries, with some of the first descriptions tracing back to Madagascar in the seventeenth century. But it was Charles Darwin who published the first scien-

above: Waterwheel plant

The amazing structure of this species of carnivorous plant, *Aldrovanda vesiculosa*, gives it its common name, the waterwheel plant.

———

tific analysis of these plants, in his 1875 book *Insectivorous Plants*. Thanks to the exotic plant trade during his time Darwin was supplied with a vast array of insect-eating plants from all over the world for him to study, focusing on one species in particular from North America called *Dionaea muscipula*. In the 1760s the Governor of North Carolina, Arthur Dobbs, had discovered this plant and given it the name flytrap sensitive plant. However, a specimen reached an English botanist called John Ellis in 1771, and – unaware of the previous description – he attributed the Latin name *Dionaea muscipula* to it, meaning Venus's mousetrap. This is the species we know today as the Venus flytrap, a plant which Darwin believed to be 'one of the most wonderful in the world'.

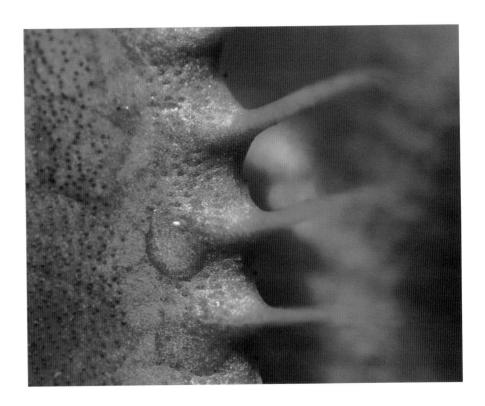

above: Sugary lure

Insects are attracted to the sweet
secretions along the edges of the
flytrap, before being swallowed
up by the plant.

—

The trap of the Venus flytrap evolved from a leaf which has been modified over millions of years into a pair of opposing pads, which can close together in the blink of an eye to capture any insect that lands on them. Insects are attracted to a nectar-like substance which is secreted from the base of the finger-like teeth which fringe the trap. Fine hairs on the surface of the trap register the movement of a landing insect, and if two of these hairs are touched within 20 seconds a biochemical signal is sent to the cells on the outside of the trap. This immediately causes these cells to fill with water, triggering cells on the inside to contract. The lightning-fast closure of the trap – in about 100 milliseconds – is one of the fastest movements in the plant kingdom. Digestive juices which are secreted from the plant's trap break down the body of the insect, releasing nitrogen which is then absorbed into the plant. Similar snapping traps are found on a rare carnivorous plant called *Aldrovanda vesiculosa*, commonly referred to as the waterwheel plant and found in the acidic lake habitats of temperate to subtropical regions. The waterwheel plant is able to capture passing invertebrates with its 3-millimetre-wide translucent green traps which when triggered can shut in just 20 milliseconds, five times faster than the Venus flytrap. The waterwheel plant is confined to just a few habitats, however, whereas other underwater carnivorous plants are far more dangerous for aquatic invertebrates – these are the bladder-worts (*Utricularia* sp.) which have adapted to grow prolifically in aquatic habitats all around the world, from the Arctic to the tropics, where they form thick mats of light green grass-like threads. The bladder-like traps of *Utricularia* were first thought to be air sacs which helped keep the thin filamentous plant afloat, but more recent studies of the genus have revealed them to be the most complex types of carnivorous trap of any plant on the planet. Inside the transparent bean-shaped trap, which can range between 0.2 millimetres and 1.2 centimetres, a vacuum is formed, and the only way in or out is via a trap-

above: Sticky death

The glistening mucilage of the sundew attracts its insect prey, which then suffocates as it is engulfed by the sugary secretion.

—

door flap with hinges on one side. The circular mouth of the trap is surrounded by branching filaments which mesmerise and guide insects towards the trap-door and also help prevent floating debris from getting sucked into the trap. Unlike the clamping jaws of the Venus flytrap, in which a reaction from the plant is required to trigger its trap, the mechanism of the *Utricularia* is purely mechanical. The imbalance of the low water pressure inside and the high water pressure outside is kept in equilibrium by a thin membrane called the velum which seals the entrance to the trap and is linked to a delicate hair-trigger. As soon as this trigger is touched by any passing organism the seal is disturbed, allowing the trap's walls to spring outwards, at the same time sucking in the column of water surrounding the creature. In less than 10 milliseconds the prey is sucked into the trap, which then closes behind it, and in a matter of hours the plant will begin to digest its captive using powerful enzymes secreted into the trap.

There are other carnivorous plants that have evolved the ability to trap insects by using sticky surfaces which act like natural fly-paper. The most strik-ing examples of this behaviour are seen in the genus *Drosera*, of which there

Perhaps the most intriguing carnivorous plant is *Nepenthes rajah,* which feeds off shrew and rat droppings.

—

are around 130 different species known to live mostly in the damp peat bogs and savannah grasslands of the world. Ranging in heights from just a few millimetres up to 30 centimetres, their carnivorous, paddle-shaped leaves which protrude from the acidic bogs are covered in numerous finger-like projections resembling tentacles. These tentacles exude a substance called mucilage that sparkles like drops of dew in the sunshine, giving them their more common name, sundew. This sticky fluid contains sugars which attract flying insects to the plant, causing them to be attached to the mucilage. Within seconds the sticky dew starts to engulf the insect, blocking its breathing pores, and once entombed inside this sugary pool it slowly suffocates. As it struggles to free itself it triggers the finger-like tentacles of the leaves to tighten the plant's grip on the animal, at the same time creating a small cavity in which it can begin to digest the insect. With the help of powerful enzymes its insides are slowly turned into a nutritious soup that the plant can absorb through the surface of its leaves.

Another dew-producing plant called *Roridula dentata* is found in the Cape Fold Mountains in South Africa. Studies of this plant's leaves by the botanist Rudolf Marloth in the early 1990s revealed that this species produces dew that is made of resin rather than mucus – the crucial difference being that mucus is water-based. The resin is the same type of substance that makes up the sap in pine trees. Marloth surmised that digestive acids and enzymes would not be able to diffuse through it in the way they can with other dew-producing plants, suggesting that although this plant can capture insects it is not actually able to absorb any nutrients from the insects trapped on its leaves. It seemed a mystery as to why the *Roridula* was able to trap insects that it could not digest. Eventually the explanation to this plant's seemingly hapless carnivory was revealed, in the form of a type of mirid bug known today as *Pameridea roridulae*. Individuals were located on the leaves of numerous *Roridula* plants living in the wild, but instead of getting stuck in the globules of sticky glue like the other insects that visited the plant, they appeared to be able to walk effortlessly along its leaves. It was found that a greasy glue-proof layer on the body and legs of the mirid bug allows them to brush past the plant's sticky parts without becoming caught, enabling them to feed on the undigested bodies of captured insets. On finding a trapped insect

'Most plants require the help of animal species for the dispersal of their seeds and the pollination of their flowers.'

they spear them with their pin-sharp mouthparts before sucking them from the inside out. Crucially for the plant, after a good feed the mirid bugs will eventually defecate onto the *Roridula*'s leaves, and in doing so provide it with a nitrogen-rich fertiliser which can be absorbed into the plant.

The Victorian collectors who first discovered many of the different capturing and killing mechanisms of the carnivorous plants we know today were fascinated by the surfeit of species which flooded into Britain in the eighteenth and nineteenth centuries. Their favourites were the Asian pitcher plants (*Nepenthes* spp.), which grow on long tendrils either high up in the trees or on the forest floors of Southeast Asia. They have large fluid-filled jug-like structures called pitchers which act as pitfall traps to capture a range of prey from tiny ants to whole rats. In 1658 Etienne de Flacourt, the then Governor of Madagascar, initially described this bizarre form of plant, but it wasn't until 130 years later that the first specimens of *Nepenthes* were introduced to the Royal Botanic Gardens at Kew by Sir Joseph Banks. Seeing the incredible nature of these carnivorous plants brought to England by Banks in 1789 ignited the European interest in the genus, culminating in what has later been described as the 'golden age of *Nepenthes*', which saw scores of newly discovered species from the East Indies fill the glasshouses of esteemed European collectors over

a period of 100 years. Most of these pitcher plants have two types of pitchers; lower ones for catching ground-dwelling prey and higher ones to catch flying prey. Depending on the species of *Nepenthes*, animals can be lured to

above: Passionflowers

The unique structures of these intricate blooms are visited by a number of pollinators including bees, bats, wasps and hummingbirds.

———

the pitcher by the presence of nectar secreted around its rim, by bright red or green patterns, or even by the bodies of rotting victims that have previously been trapped. The lip of many pitchers is covered in a slippery substance, and the fluid in the body of the pitcher is a viscous soup of digestive enzymes. In some cases it is strong enough to dissolve a piece of steak in days. The liquid has been found to be made of viscoelastic filaments which act like quicksand – the more the insect moves and struggles to free itself the more trapped it will become. Epicuticular wax crystals coat the sides of the upper pitcher wall, so it impossible for insects to grip with their feet. Trapped animals are slowly broken down and the released nutrients which are dissolved in the digestive fluid are absorbed through multicellular glands on the inner pitcher wall.

As well as consuming the animals that they trap in their fluid-filled pitchers, many *Nepenthes* also provide protection or shelter for certain animals, and much like the bromeliads of South America some of them can support their own mini-ecosystems: mosquito larvae can live in the fluid of the pitchers, frogs lay their eggs in some species, and in others bats have even been found to roost. In the lowland peat forests of northwest Borneo one of the largest and most spectacular of the genus, *Nepenthes bicalcarata* (fanged pitcher), accommodates a certain species of carpenter ant called *Camponotus schmitzi*. Hidden under the rim of its huge pitchers – which can reach up to 25 centimetres in length and 16 centimetres in diameter, giving them a volume of over a litre – is an unseen army of ants. But unlike the ants which aggressively patrol the spines of some cacti these ants remain concealed. Until, that is, a large insect such as a cockroach or a cricket falls into the pitcher. As the insect slowly succumbs, one of the ants leaps into the fluid, taking hold of its body with its powerful pinchers. As the insect is retrieved the other ants help pull it up out of the fluid. It is a slow and steady process, but with the help of the hook-like tarsal claws on their feet a group of ants is able to pull an insect many times their own size up the pitcher's steep slippery wall. Once back in the safety of the lip of the pitcher the ants set about dismembering their catch, feasting on the best bits of the insect and throwing the rest back into the plant.

Researchers who observed this behaviour were confused as to why the plant permitted the ants to steal its insect prey, questioning why it hadn't developed a stronger digestive fluid or ant-proof slippery sides. Eventually a study of the amount of nutrients that the plant is able to absorb with and without ants provided the answer – it was found that the pitcher could not digest large insects, and if left in the fluid they would rot and become stagnant. By removing them and dropping just small pieces back into the fluid the ants essentially prevent the plant from overeating, thus being beneficial to it.

Where the fanged pitcher employs animal assistance to remove excess food from its traps, the endangered *Nepenthes rajah* – which grows in open areas of grass between 1500 and 2600 metres on just two mountain peaks in Borneo – attracts animals to drop food into its traps, which then leave unharmed. With

opposite: Unwelcome guest

Longwing butterflies such as this *Heleconius erato* have evolved a unique but somewhat uneasy relationship with *Passiflora*.

—

Extreme Survival

Chapter
Seven

'Unlike the soils of many moist habitats, where water and nutrients are in plentiful supply, the soil of deserts lacks organic matter.'

Life in the Extreme

The world's arid environments, which comprise the planet's desert biome, are so extreme that the plants that survive in them are some of the most highly evolved on the planet. Though plants have their origins in the ancient lakes and oceans of the Earth, innumerable adaptive steps have enabled some to become modified to a life further away from their aquatic beginnings, over time colonising drier and more inhospitable lands.

Since the rise of plants in Earth's oceans over a billion years ago the climate has fluctuated through prolonged cold glacial periods interspersed with warmer interglacial periods. The most recent period of mild temperature has roughly spanned the last 10,000 years of human civilisation. But contrary to this short- term trend of warming, on a broader timescale of the last 60 million years the planet has seen a pattern of increased levels of glaciations. As a result, huge areas which were once green and tropical have become more barren, arid landscapes. In response to the increasingly arid climate, plants from previously humid habitats evolved to cope with extreme climates. For example, a particular jungle thorn bush called *Pereskia* developed many characteristics by which modern cacti are defined today – long spikes emanating from its tall woody stem, thickened waxy leaves to store water, and the ability to respire using a chemical pathway called crassulacean acid metabolism (CAM photosynthesis), which allows the plant to open its stomata to absorb carbon dioxide only at night, therefore minimising water loss. This plant first evolved roughly 50 million years ago, and it is believed to be the ancestor of all cacti, perhaps the most extreme group of dry-zone plants adapted to life in the desert.

Over time many new drought-tolerant plants radiated, each possessing different adaptations to life with little water. Plants from the family Aizoaceae or ice plants evolved to live in the deserts of southern Africa; they adapted to grow largely underground to stay cool and minimise water loss, and they developed their strategies of pollination and reproduction to suit the seasonally scorched landscape. A family of plants called Didiereaceae which are endemic to Madagascar developed thickened water-storing structures in their leaves and tall thin stems. Another family of plants from the arid regions of the southern hemisphere called Portulacaceae developed thick leaves and stems to retain water, while the Cactaceae family from the Americas evolved into some of the most iconic desert plants known today. These dry-zone plant families diversified over the past 40 million years, and as modern continental positions and climates have become more stabilised over the last five million years, they evolved into the plant species which fill the Earth's arid habitats today.

Increasing ice deposits in Antarctica as well as the formation of mountain ranges caused further aridification of certain regions of the planet. The colossal uplift of the Andes formed a rain-shadow along the spine of South America and ice build-up at the South Pole caused the creation of a cold offshore Peruvian current. Today the Atacama and Sechura deserts of Peru form the driest landscape on Earth, including one of the few places where rain has never been recorded. The same effect was seen in the northern hemisphere, where the formation of the Rocky Mountains blocked coastal moisture from reaching the mid-continent, starving the modern Great Plains and the Mexican Plateau of moisture. Similarly the uplift of the Tibetan Plateau led to the desertification of large parts of Asia and is believed to have contributed to the increasing aridity of the Sahara.

Today deserts make up about a third of the surface of the planet, comprising an area of around 80 million square kilometres. They are characterised by their extreme temperatures and near-absence of precipitation, which is usually less than 20 centimetres a year, compared to the 200 centimetres that fall on the Amazon basin. The dry winds which shape the surface of the landscape like sandpaper increase the aridity further still. Unlike the soils of many moist habitats, where water and nutrients are in plentiful supply, the soil of deserts lacks organic matter. In some cases the environment for survival may be so limited that many of the plants that live there stand alone in order to reduce competition between them. However, although these desert areas appear to be barren wastelands compared to tropical forests, they are in fact biologically rich habitats.

Each desert environment around the world is defined by its own plant life. The Namib Desert of southern Africa is believed to be the world's oldest continuous desert, dating back to around 55 million years ago. Today it contains a number of plant species which have evolved unique traits in order

above: Babies' toes

These plants from the Aizoaceae or ice plant family store water in their thickened leaves.

—

to survive in this relentless climate. *Welwitschia mirabilis* can live for many thousands of years, existing as no more than a pile of dried, brown leaves. *Pachypodium namaquanum* – or elephant's trunk – grows as a 4-metre-tall, bare totem-pole topped with a few fleshy leaves, and quiver trees (*Aloe dichotoma*) rise out of the gravel scrub resembling a giant bouquet of miniature palm trees supported by a main trunk wrapped in sharp bark. In America's Sonoran

Desert saguaro cacti dominate the landscape with their 21-metre-tall arms enshrouded in a cloak of spines, viciously spined agaves store water in their succulent leaves, and a sea of vivid brittle-bush flowers flood the desert floor from March to June. The expansive sunburst red sands of Western Australia cover just under half of the continent and are punctuated with short scrubby tufts of spinifex grass, saltbush shrubs, and yellow flowers of the nation's emblematic wattle trees.

Whether it is hot and dry or cold and dry, a plant's primary problem in these dry zones is coping with lack of water. For large parts of the year these environments are either baked by the sun's rays, or they are whipped by sub-zero winds which lock up any moisture in ice. The waxy cuticles of most vascular plants act as a barrier to prevent water loss in temperate climates. When these plants lose a small amount of water from their leaves via transpiration it is soon replaced by the process of capillary action which eventually wicks more water up from the plant's roots. However, in hot arid habitats where water is not readily available in the substrate surrounding a plant's roots, there is nothing to replace lost water. Dry-zone plants have evolved numerous mechanisms and structural modifications to deal with the scarcity of water, both in minimising water loss and in developing bodies which can store water. Collectively these drought-tolerant plants are called xerophytes, more often collectively known as cacti and succulents.

Some desert areas may never see rain at all, but most deserts do at least receive a few centimetres of precipitation a year. This typically falls in sudden torrential downpours, so the key to plant survival is to store up as much water as possible. The first step taken by many xerophytes to achieve this was to evolve succulent water-storing body parts in the form of thickened structures composed of multiple layers of cells. The leaves of *Echeveria laui* from Oaxaca in Mexico have evolved to become water-storing parts, resembling swollen pink and white paddles. The cacti from the Americas have reduced their leaves to become nothing more than spines, to prevent water loss, and their stems have evolved to be able to photosynthesise as well as store water. The golden barrel cactus (*Echinocactus grusonii*) from Mexico's Moctezuma River canyon, whose spherical form is the ideal water-storing shape, minimises water loss through transpiration. Many cacti, together with the euphorbias, their Old World equivalents from Africa, are also covered in a thick flexible skin with pleats which run down the length of their bodies, allowing them to expand and shrink like a concertina as they absorb and use up water throughout the year. This thick skin also insulates the plant from searing ground temperatures, which can reach as high as 60°C, and prevents their stored water from evaporating into the atmosphere.

The caudiciforms have specialised roots or bases in which they can hold large quantities of water. These can range from small but beautiful spindly growths of the elephant's foot (*Dioscorea elephantipes*), whose deeply fissured root mass could be mistaken for a pile of chopped wood, to baobabs (*Adansonia digitata*), which grow in the savannahs of sub-Saharan Africa. The baobabs are often referred to as upside-down trees, as their relatively short and fat upper parts look as if they should be the tree's roots, not its branches.

While cacti and other succulents are adept at absorbing and storing large amounts of water, they still need to open up the stomata in their thick skin to absorb the carbon dioxide necessary for photosynthesis, and this makes them vulnerable to water loss. To avoid opening up their stomata in the heat of the day, many dry-zone plants have evolved a method of opening them only at night, when evaporation rates are low. However, as they are unable to photosynthesise in the dark they have to store up the absorbed carbon dioxide until the morning. They do this by converting it into a chemical called malic acid. The following day this acid is broken down to release the carbon dioxide, and in this way the plant is able to photosynthesise with its stomata closed. Through this mechanism, known as crassulacean acid metabolism (CAM), desert plants lose 90 per cent less water per unit of carbohydrate synthesised than a non-CAM plant. As well as being super-efficient at retaining moisture, CAM plants also have the added advantage of being able to shut down their metabolism during times of extreme drought. When water is in short supply the plant's stomata can remain closed permanently both day and night. The moist environment inside its cells allows a very low level of metabolism to keep

going using nocturnally fixed carbon dioxide. The plants can remain in this state until water returns to their habitat, and they are able to regain their normal metabolic rate in just 24 hours.

For the short time that water is available it is necessary for desert plants to have a highly adapted root system to absorb as much water as possible before it evaporates. Where plants in the tropics and temperate regions have long roots designed to reach deep into the ground to absorb stored water, plants in arid habitats have relatively shallow roots which sit just under the surface to suck up rain as soon as it falls. While cacti roots are not deep they are still extensive, and a metre-tall plant can have roots which extend out 3.5 metres horizontally – on average the roots of most dry-zone plants cover an area up to twice that of their canopy above ground. Cactus roots are covered in a springy cork-like coating which prevents them from losing water, and over the rainy season they put out many new root hairs to increase water uptake. After rainfall, when the ground begins to dry and crack again, sections of the roots die off – a small sacrifice to prevent the plant as a whole losing too much water. Many plants also drop their leaves during periods of extreme drought. To prevent water being lost via transpiration through their leaves, water is first drawn back into the plant's stem and the leaves simply dry up and fall off. The creosote bush which grows in sporadic tufts in America's Sonoran, Chihuahuan and Mojave deserts is unable to drop its leaves but instead covers the surface of its leaves with a thick resin which seals in the moisture.

However, some of the most extreme dry-zone plants, which often have to endure up to a year without water, don't attempt to resist water loss but instead have evolved the capability to almost completely desiccate. One such plant is the small *Selaginella lepidophylla*, or resurrection plant, that grows natively in Mexico's Chihuahuan Desert. This primitive plant is a lycopod, a small moisture-loving plant. During a period of drought it becomes a shrivelled brown ball of dried leaves no larger than an orange, and its metabolism slows almost to a stop. In this shrunken form its surface area is drastically reduced so that some moisture may be retained, locked in the centre of the ball. The plant lies dormant, often staying in this state of suspended animation for many years.

(a)

(b)

Eventually the ground surrounding the resurrection plant will begin to saturate with rejuvenating drops of desert rain. As water is absorbed by its dried body, the plant's metabolism restarts. Within a few hours the brown leaves begin to unfurl, plumping up as they fill with water, and fan out into a broad rosette of dark green branches 30 centimetres across. Its giant lycopod relatives were the dominant land plants 300 million years ago, but they have since gone extinct, and only the remarkable resurrection plant lives on.

The lycopods to which *Selaginella lepidophylla* belongs are the oldest living vascular plants, not dissimilar to the first plants that grew in the coastal habitats of the Silurian. *Selaginella* does not produce complex flowers, or seeds or pollen – instead, it produces spores – and the mesophyll layer of its leaves is only a few cells thick.

An angiosperm which is equally able to 'resurrect' is *Blossfeldia liliputiana* – a nod to the race of people encountered by Gulliver in Swift's classic tale and the smallest species of any cactus. This tiny plant, whose grey-green stems reach no more than 2 centimetres in diameter, is fairly widespread in the arid habitats of the east side of the Andes, including southern Bolivia and southwest Argentina. It is wedged into cracks between rocks and when it dries up its swollen body shrinks and becomes flattened like a disk. Unlike all other cacti *Blossfeldia* almost completely lacks stomata, but instead has breathing pores called areolar pits sunk deep into little depressions across its surface. It is from these structures that the spines of cacti grow. It is believed that *Blossfeldia* has the fewest stomata of all photosynthesising plants on the planet, a trait

which has clearly evolved in response to the extreme life strategy of this plant. Another trait unique to *Blossfeldia* is that it has no thickened outer cell wall and so it does not retain water like other cacti, but instead it allows its body to completely desiccate. In its dehydrated state this plant can wait for many months. But when rain does eventually return it can refill its cells and plump up its body once more, and within a matter of weeks it can produce beautifully delicate white flowers.

For the rest of the plant world, whose cells would simply shrivel and irreversibly deform if they allowed their bodies to dry out, the key to staying alive in extreme heat and drought is to become a master at retaining precious water stored in specialised body parts. Some of the most effective examples of such adaptations are best exemplified in the cactus family which – aside from one species, the mistletoe cactus from Asia and Africa – all originate from the deserts of the Americas. Acting similarly to the expandable ribs seen on the golden barrel cactus, many cacti such as *Thelocactus* from the scrublands of Mexico have structures called tubercles which are cone-shaped protrusions covering the plant's surface, allowing it to expand and shrink without bursting its skin. Another Mexican species, *Stenocactus crispatus,* is covered in a deeply ridged sinuous ribbing which gives it the appearance of a strange green coral. Both the tubercles and the wave-like ridges increase the cacti's surface area for absorbing sunlight without overly increasing the risk of water loss. In species where the ribs are most pronounced – creating 5-centimetre-deep pleats down the entire cactus's body – this also provides the plant with a way to reduce the heat of direct sun, as when the sun shines from any side the plant's ridges will mutually shade the rest of the plant. Another way for a cactus to reduce the glare of the sun is to reflect it away from its stem, and some cacti achieve this by covering themselves in a powdery layer of sunscreen, such as *Pilosocereus pachycladus,* whose body is covered in a stunning azure blue coating.

One way for a plant to reduce the amount of sun that it absorbs is to escape it completely, opting for a life underground, or at least partially so. This semi-geophytic existence helps keep the plant cool and in doing so reduces the amount of surface exposed to the drying rays of the sun. Some examples include the slow-growing spineless cacti from the genus *Ariocarpus* that have evolved to live with only the tips of their fleshy star-shaped tubercles above the ground, while the rest of their body sits buried in the soil. Another succulent plant called

Fenestraria – or window plant – from the ice plant family has evolved bulbous green tube-shaped growths which sit in the substrate with only their tops showing. At the top of each tube-like growth is a flat area of translucent tissue which works much like a camera lens, moderating the amount of sunlight which the plant absorbs for photosynthesis. In the deserts of Brazil and Chile the sun is so powerful that it can even penetrate the translucent quartz rocks which cover the ground; using this cover to their advantage, the endangered *Discocactus horstii* can be completely covered by these rocks and still absorb enough solar energy to effectively photosynthesise, often only revealing itself above ground when it puts out its bristly white flowers to attract hawk moths.

The impregnable covering of sharp spines that have evolved from reduced leaves which enshroud most cacti are often wrongly assumed to be purely for protection, but in fact they also play an important role in controlling the temperature and moisture content of the plant. In species such as *Pachycereus schottii*, often known as the whisker cactus, the spikes grow tightly packed together, forming an insulating mat which helps shade the stem and traps humidity close to the plant to reduce water loss by transpiration. In many species of *Mammillaria* cacti this dense spination has evolved even further to create tufts of woolly hairs which surround the plant's areoles creating a moist environment close to the plant's stem. These spiny hairs also help protect the delicate flowering parts from drying out and from hungry predators when they are in bloom. In the most extreme examples dense mats of spines grow so thickly over the stem of a cactus that it creates its own microclimate underneath. Even in the heat of the day this keeps a layer of moist and relatively cool air around the plant, creating an efficient boundary against water loss. In some deserts, where the only source of water is in the form of a coastal fog which drifts on the breeze, cactus spines can act as a fog trap. Under the magnification of a scanning electron microscope one can see that the surface of the needle-like spikes of the *Eriosyce paucicostata* cactus from Chile are not smooth but actually covered in fine ridges and rough channels. Their coarse surface collects fine droplets of water from the air, and as they build up they drop on the ground below to be absorbed by the plant's shallow roots. Some cacti, such as the *Eulychnia* species, even have lichens growing on their long thin spikes which help them to collect moisture from fog. Along the coastal parts of the Atacama

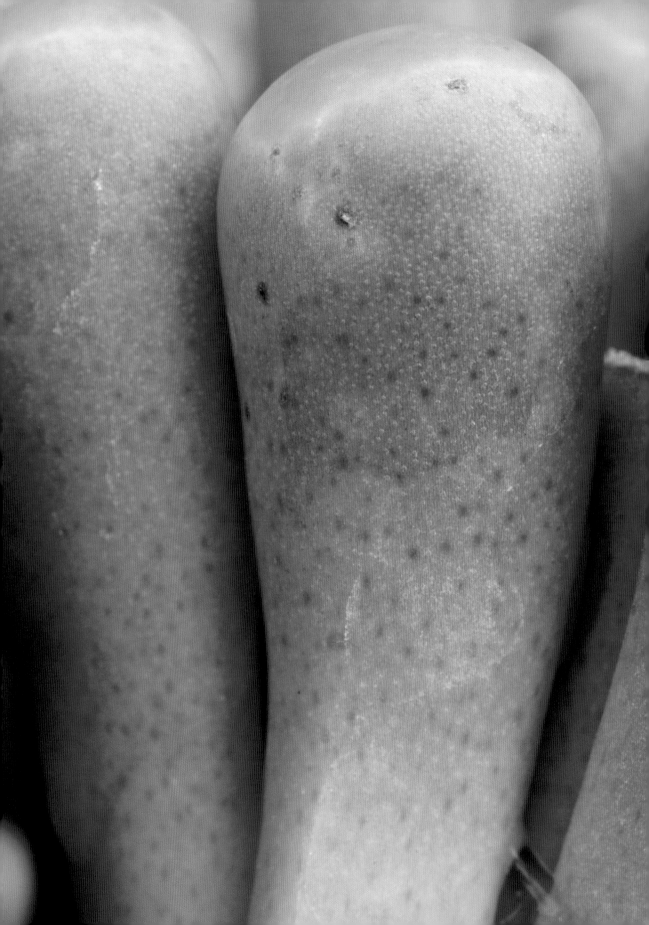

opposite: Window plant

The tips of this plant's succulent leaves are made of translucent tissue acting a bit like a lens to regulate the amount of light it absorbs.

—

Desert it is not uncommon to find these tall columnar cacti almost completely enveloped in a thick cloak of lichens. Some cacti growing in the Atacama have rough waxy structures on their surface which trap droplets of fog and channel them down towards the base of the plant; plants like the star-shaped *Astrophytum myriostigma* and the sea-urchin-like *Copiapoa cinerea* cover their entire bodies with this texture, invisibly collecting water in a land which appears to have none.

The water-capturing and water-storing capabilities of xerophytic plants – in an environment largely devoid of water – makes them an obvious target for any thirsty creature which resides in the dry zone. Desert plants provide a whole array of herbivores with food and water. Consequently over millions of years of evolution these plants have developed a number of defence mechanisms. The most subtle of the dry-zone plants are small enough that with suitable camouflage they can grow unnoticed by passing herbivores. This is the strategy adopted by a group of small ice plants called *Lithops*, which grow widely from sea level up into the mountains in South Africa and Namibia. With a name derived from the Greek *lithos*, meaning 'stone', and *ops*, meaning 'face', these short globular plants grow as two fleshy leaves fused at the base. Growing together in clumps, their flat pebble-like appearance allows them to blend in with the surrounding rocks to avoid being eaten. *Lithops* are a highly desirable plant for collectors all around the world, and the multitude of different colours and patterns which they exhibit correlates with the various substrates in which they are camouflaged: blue-grey leaves blend in among quartz rocks, mottled green and black leaves are cryptic in gravel, orange and brown patterns help them remain hidden in sandy ground, and near-white growths keep the plants concealed in saltpans. It is often only during the period of summer rainfall that these plants become apparent in their landscape, when each pair of tender leaves splits apart to produce a single delicate flower.

Some of the larger desert plants also use forms of disguise to keep themselves hidden from herbivores, and a handful of cactus species ingeniously avoid being eaten by grazing mammals by looking like something that may have already been eaten – they resemble droppings. Herds of hungry llama-like guanacos (*Lama guanicoe*) together with javelina (*Pecari tajacu*) and an introduced population of goats are a menace to the plants of Chile, stripping their succulent stems for a quick meal. But one species of cactus called *Copiapoa laui* has evolved a way to make it look especially unappetising, by growing in dark patches of spineless, flat, diskette growths roughly a centimetre in diameter, which to the untrained eye seem to be a scattering of mammal dung. Another plant which uses the same method is a species of cactus called *Lophophora williamsii* found in the Chihuahuan Desert. Its 7-centimetre-wide stems often become flattened to only a few centimetres and as they grow and overlap each other they take on the distinct shape of a large pile of mammal droppings. However, because of its size *L. williamsii* has had to evolve a more powerful defence. Its squat stems are full of highly toxic chemicals, containing upwards of 60 different alkaloids, and as a result it is extremely bitter and distasteful to herbivores. This dissuades any animals from attempting to eat it. Should any human ingest the plant's powerful toxins, though, the effect is more extreme, as the principal alkaloid is mescaline, a powerful psychoactive drug. The potent hallucinogenic properties of *L. williamsii* and its relatives appear to have been known for thousands of years, with archaeological evidence from Texas suggesting that it was used by the native people of America's southwest as far back as the middle of the Archaic period, around 5200 years ago. Referring to the plant as peyote, a derivative of the Nahuatl word *peyotel* or *peiotl*, Native American tribes heralded its curative properties for diabetes, fever and the relief of pain during childbirth. The Huichol people of Mexico regard the 'sacred peyote' as a gift from god, and the overwhelming spiritual images evoked by the hallucinogenic effects of the cactus have had a profound effect on their religious beliefs and philosophies. In more recent history the powerful narcotic effects of *L. williamsii*'s chemicals are accredited as having influenced a whole generation of musicians, writers and poets, ranging from Allen Ginsberg to Aldous Huxley.

The most obvious method of protection used by cacti and succulents living in arid environments is their impenetrable shield of spikes, spines, thorns or prickles. These sharp woody growths have evolved as a result of adaptations towards smaller leaves to reduce water loss – as smaller leaves lose less water than larger ones, smaller leaves were selected for over millions of years of evolution, and as they diminished they also became sharp and pointy. Gradually they became saturated with calcium carbonate and pectin, which made them rigid and tough. Eventually the thickened stems of cacti took on the photosynthetic role of their leaves, allowing the leaves to become modified into the needle-sharp spines we see today. Evidence for this gradual transition can be found by looking at the earliest stages of developing cacti spines, which look nearly identical to the earliest developmental stages of leaves on other plants. They are produced at the base of the axillary bud's shoot apical meristem. However, apart from their shared origins cacti spikes bear no resemblance to leaves whatsoever: they have no xylem, no phloem, no stomata and no chloroplast, and when they are mature their cells are dead.

above: Desert survivor

While *Welwitschia mirabilis* rarely look very healthy, dried and twisted on the desert floor, they are known to live for up to 2000 years.

—

There are a great number of spiky desert plants which thrive in arid habitats all across Africa, but the spikes of these plants have evolved differently. For example, *Euphorbia neohumbertii* from Madagascar exhibits a body very similar to the cacti of America as a result of evolving under similar environmental conditions – it has a tall, water-storing green body, covered in spikes. To the untrained eye this plant looks just like a cactus. But the spikes of *Euphorbia* and their relatives have evolved from modified shoots, not leaves, and where cactus spikes are smooth and grow in unbranched clusters, the spikes of *Euphorbia* only occur singularly and never in clusters, often branching with tiny scale-like leaves. Where a cactus is essentially a water-storing stem with no leaves, the doppelganger *Euphorbia neohumbertii* is a giant water-storing leaf with nearly no stem. The amazingly similar appearances of these unrelated plants, which have evolved independently thousands of kilometres apart, is the result of a phenomenon called convergent evolution, and it is seen in a number of other dry-zone plants which have adapted to similar environmental constraints in separate habitats. For example, aloes from Africa and agaves from South America have both evolved tough fibrous spear-like leaves arranged in a rosette. Aizoaceae from southern Africa and *Ariocarpus* cacti from Mexico have both evolved to bury their small succulent bodies in the substrate to keep cool. Similarly, *Euphorbia obesa* from the Eastern Cape and *Astrophytum asterias* from Texas have evolved seemingly identical octagonal/round bodies.

above: Drinking fog

This species of *Astrophytum* or star cactus has a rough waxy surface which captures minute droplets of airborne water which are then channelled by its ridges to its base.

—

In whatever way they are produced, the protective spikes of dry-zone plants appear in a dizzying array of forms, appearing in innumerable variations of shape, number and function. Some cacti, such as *Mammillaria poselgeri*, known as the fishhook cactus, have long hook-shaped spines which are thought to 'educate' herbivores not to attempt to eat the plant after an initial 'hooking' by the plant's spines. The same effect is achieved by the pairs of opposite-facing thorns of *Acacia mellifera*, which catch the fur of grazing animals that get too close to it. Some of the longest spikes can grow up to 25 centimetres long, as seen on *Trichocereus chiloensis* from Chile, and the *Tephrocactus paediophilus* has razor-sharp, flimsy spines two or three times the length of the plant. More often spines grow to just a few centimetres in length, protruding in crown-like bundles from the plant's stem. These bundles can be long and thin or can produce stocky, broad clumps of spikes. Some more primitive cacti have spines which cover their flowers as an added defence, and other cacti have adapted their spikes to have an added advantage of camouflage such as *Sclerocactus papyracanthus* or grama grass cactus, whose sharp 5-centimetre-long spines have evolved to look like dried brown grass to conceal its succulent green body from herbivores. Cacti also have highly effective, almost microscopic, spines called glochids, commonly seen on species of *Opuntia* or prickly pears. At microscopic levels these tiny hairs can be seen to have barbed tips, making them particularly tricky to remove.

While the drought-resistant plants of the world's arid landscapes are masters at retaining their water and keeping away unwanted animals, there are times when these unmovable desert dwellers must leave themselves susceptible to desiccation to produce the flowers which attract their pollinators. For all xerophytes the production of flowers is a costly venture, as flowers are relatively delicate structures prone to drying up and losing water in extreme heat. Where the flowering plants of the temperate and tropical regions of the world can produce blooms all year round, those in the dry zone must be more conservative with their resources. Instead of having long flowering periods, it is far more economical for them to wait and only produce their flowers when conditions are optimum. This ability to wait for the right conditions enables many plants to survive in the dry zone.

In true deserts like the Tucson Desert the opportune time for flowering can be as infrequent as every 10 years, and only after substantial winter rains and an amiable summer climate is the environment transformed by a sea of floral colour. When these conditions arise, carpets of yellow and orange Californian poppies (*Eschscholzia californica*), spearhead-like lilac lupines and the deep-purple heads of owl clover (*Castilleja exserta*) bring a stunning spectrum of colour to the desert. The same is seen every April in the semi-arid lowland steppe of western Kazakhstan, where a small amount of precipitation sees the dried brown ground turn a lush green, as the landscape is filled with a host of tulips in a dazzling array of yellows, purples, whites and reds. But after only a few months of colour, these pastures are once again turned barren by dry winds.

Hylocereus undatus, the 'queen of the night' cactus from Central America, flowers just one night a year. The shrivelled appearance of this vining cactus is a tangled mass of succulent tubes, each a few centimetres in diameter, that tumble over rocks and hang over the branches of other plants on which it has

become established. It is, however, perhaps the most exquisite member of the cactus family. As the temperatures in the tropical deciduous forests where it grows can reach a crippling 40°C in the summer, this cactus is

opposite: Desert colour

The springtime wildflower bloom of Antelope Valley Californian Poppy Reserve is one of the most spectacular desert sights.

—

largely inactive during the day. In late spring each year its sprawling vines begin to produce small protrusions which grow in length over a number of weeks, and by midsummer these have become 20-centimetre-long spear-like green buds. The plant then sits idle, remaining poised for a full moon, when it springs to life in a display of astounding beauty. As the sun begins to set and the light of the moon emerges, the flower's buds begin to move, slowly extending the sticky ends of its yellow stamen from the tips of its buds, closely followed by a brush of pollen-covered anthers. Over the next couple of hours the fleshy green clasp-like tepals and bracts begin to peel back as the bud slowly opens to reveal a head of creamy white petals, unfurling into a glorious flower. As the flowers open they emit a sweet fragrance, attracting hawk moths and nectarivorous bats that fly up the odour plume. As they get closer to the plant the magnificent 30-centimetre-wide flowers appear to glow in the light of the moon, drawing its pollinators to feed on its nectar. After only a few hours, the flowers of this night-blooming cactus begin to close again, and by the time the morning sun-light emerges, the flowers have already wilted and died. This is all part of the plant's survival strategy, however, as the closed petals form a protective shield around the newly fertilised seeds, locking vital moisture inside. The queen of the night's approach to producing flowers that only last one night is a tactic which has proved highly successful.

Agave americana is more commonly known as the century plant, as those who first discovered it in its native habitat in Mexico believed it to flower only once every 100 years. The name has stuck, even though it is now known that it does not flower as infrequently as that. The powdery blue leaves of this massive succulent are edged with serrated tooth-like structures which clump together to form a wide rosette reaching 2 metres high and 4 metres wide. Like many plants that have evolved to survive the harsh conditions of the dry zone, *A. americana* is very slow-growing, and as it grows it builds up food stores of sugar and starch in its tough body. As the plants that share its habitat go about

above: Queen of
the night

The dinner-plate-sized flowers of this
plant are beautifully fragrant but only
bloom for one night a year.

—

their regular cycles of flowering and dying,
the agave remains unchanged. Over a passage
of 30 years or more, the agave will slowly
expand as it gradually builds its internal store
of energy.

Eventually the agave will have amassed enough energy over its long life,
sometimes as long as 60 years, at which point it begins to channel its store
of carbohydrates into developing a flower spike. From a life of veritable stasis
the plant's metabolism kicks into overdrive, and as this spike emerges out of
the centre of its broad leaves it reaches a growth rate of up to 25 centimetres
a day. Resembling a giant asparagus stalk, the agave's flower spike rockets up
to a height of 8 metres, where it then splits to produce a metre-wide branched
structure covered in dense clusters of buds. Over a period of two weeks the
flower spike will produce tens of hundreds of strongly scented pale yellow
flowers, which act as a beacon to night-foraging bats. Each bud of the agave's
bloom is oozing with copious amounts of nectar, which is produced using the

plant's lifetime store of energy, which in turn ensures it is visited by a sufficient number of pollinators. However, it is only after *A. americana* has flowered that the importance of its patience of many decades becomes clear. When its flowering head has produced its many thousands of wind-dispersed seeds and its flower stalk collapses, so too the rest of the plant begins to shrivel. As quickly as the plant shot out its towering flower stalk to spread its genes, its metabolism slows down and the plant begins to wither and die. The plant literally flowers itself to death, yet relies on at least one of its thousands of seeds to successfully germinate.

Pushed to the Extreme

For the many varieties of plants that have evolved specialist adaptations to live and reproduce in the unforgiving conditions of the planet's dry zones, it is a result of hundreds of thousands of years of natural selection that has finely tuned them to be able to do so. As the conditions of these habitats became more extreme for plants to survive there, so the bodies of plants became more extreme. The actions of humankind, however, have had a historical propensity to upset the balance of nature, and as a result many plants now have an 'anthropogenic extreme' subjected upon them.

Such plants are perhaps worst affected on islands – areas of limited space in which the ever-growing human population is forced to strip the native vegetation for food and resources, facilitating further population growth, which in turn requires even more resources. Island species are especially vulnerable in that they have evolved in line with the very specific parameters of their unique habitat, which can easily be upset by sudden change, throwing the island ecosystem out of equilibrium. When island species are threatened by changes in their immediate habitat, they have nowhere to go. For millions of years the dodo lived a happy existence on the island of Mauritius in the Indian Ocean; its population was kept in check by the natural forces of competition and disease. However, with the arrival of human settlers in the early 1600s, the birds were unable to escape the cats and dogs brought to the island, and they were unable to protect their ground nests from the introduced pigs and macaques. The dodo was soon wiped out. This is the age-old tale which currently threatens the exotic biodiversity of Madagascar

and Borneo, the islands of Indonesia and the Philippines, as well as similar areas of rich diversity across the globe. From rainforests to mountain ranges, evidence suggests that species across the world are disappearing in variety and number faster than at any time in the Earth's recent history. But fortunately there are those who have dedicated their lives to the meticulous study of the plants and animals of almost every environment on Earth, providing an invaluable resource by which we can monitor the state of our planet.

One such resource is the Herbarium of the Royal Botanic Gardens at Kew. Since its inception as a botanical garden in 1759 Kew has been at the forefront of plant science and discovery, and the Herbarium is a testament to this legacy. Founded in 1853, the building was intended to hold the dried collections of preserved plant and fungus specimens that had been collected by botanists and horticulturalists, from both Britain and overseas. With the rapid expansion of the British Empire during the nineteenth century, the Herbarium's collections rapidly grew. Today staff process over 50,000 specimens a year that come from across the globe, with the collection as a whole now totalling over seven million individual specimens. The Herbarium as it stands is the largest repository of botanical data in the world, and more importantly, it is a key weapon in safeguarding the future of plants on our planet. This unique databank of preserved specimens along with detailed descriptions of their habitat and relative abundance plays a vital role in monitoring the global health of plants.

Exactly how important this collection is as a tool for global conservation is exemplified in the case of one particular plant from the island of Rodrigues in the Mascarene Archipelago, called *Ramosmania rodriguesii* or café marron, a wild member of the coffee family. Through their many thousands of years of geographic isolation the Mascarene islands, including Mauritius and Rodrigues, have evolved many unique and fascinating species of flora and fauna. However, since the arrival of European settlers in 1638 their endemic flora has slowly diminished in the face of habitat destruction and the arrival of introduced species. On the smallest of the islands, Rodrigues, eight species of plant are already known to have gone extinct, and of 38 remaining endemic species 21 are listed as being endangered, including one particularly enigmatic species

called the café marron. This tree, 2–4 metres tall, with lush waxy leaves and pentamerous white flowers, was first discovered in 1874, when its 'type' or primary specimen was sent to the Herbarium at Kew and it was given its Latin name, preserved and catalogued. Apart from a rough drawing of the

above: The Herbarium at Kew

The world's largest repository of botanical data, and a key resource in plant conservation.

—

plant made by a European visitor to the island in 1877, little more was heard or seen of this species for many decades, and as the populations of introduced pigs and goats increased, the plant's population began to dwindle. By the mid twentieth century café marron was assumed extinct from the island. But then in 1979 a young boy named Hedley Manan, from a school party that had been encouraged to look for a number of rare plants on the island, found a lone specimen of what appeared to be the plant from the 1877 sketch. A cutting was taken from the shrub and sent to Kew, where it was compared with the preserved specimen in the Herbarium – and indeed it was the very same species. This was great news for botanists and conservationists alike as it spelled hope for the chances of re-establishing café marron on the island.

opposite: The living dead

Café marron's story of survival is
an example of just how important
centres like Kew are for preserving
our planet's species.

—

However, with the fate of the whole species resting in this one lone plant, scientists and horticulturalists knew they would need to act quickly. With help from the IUCN and the Mauritian Forestry Service a further cutting was sent to Kew, where it was successfully propagated in the Temperate Nursery, producing dozens of genetically identical new plants.

Back in the UK the cuttings grew rapidly, and soon the plants began to produce white flowers. But despite numerous attempts it proved impossible to pollinate the flowers and therefore the plants couldn't produce any seeds. The horticulturalists at Kew began to assume the worst, with all the evidence suggesting that the last wild plant from which the cuttings had been taken must be sterile. As only one living plant was known, the horticulturalists had no other suitable specimens to compare it to, and as a result it was impossible for them to assess whether their plant was male or female, or indeed both, or if it was mutated in some way that might explain why it could not be pollinated. For many years Kew continued to propagate cuttings from the one remaining plant in an attempt to produce seeds, to no avail. In 2001 a handful of cuttings were repatriated to Rodrigues to be planted in a fenced-off area, but ultimately their population was doomed if they were unable to naturally reproduce. Before long the species acquired the name 'the living dead', as even though its cuttings could be planted and re-planted to produce an indefinite number of identical clones, without cross-pollination the gene pool of the species would be so small that a single bacterial or viral disease could wipe out the entire population. As well as the scientific and botanical struggle to unravel the mysteries of the café marron in England, the lone survivor and its re-planted clones back in Rodrigues were dealing with further persecutions still – the attention that these meagre cuttings were receiving in their natural habitat from conservationists led locals on the island to believe that the plant possessed curative properties. Fences were erected to protect the plants, but people soon cut through to cut off parts of the plants to use in hangover cures or to treat venereal diseases. Eventually three layers of 3-metre-high fences were put in place to keep the last café marron safe, and there it remained, safely entombed, waiting for someone to provide the key to its survival.

Back at Kew the continuous blooming of the cloned plants in the nurseries inspired one horticulturalist named Carlos Magdalena to try any method that could conceivably work to pollinate one of the flowers. Together with his supervisor Viswambharan Sarasan, Carlos tried various methods to cross-pollinate the flowers of the cloned plants, and eventually they came to amputating the stigma from one flower and directly transferring its pollen to another flower. Nine hundred and ninety-nine of these attempts proved fruitless, but in the summer of 2003 one flower on one particular plant showed a small swelling of its ovary, and within a few weeks this plant produced a fruit containing the crucial seeds of a new generation of café marron. Upon ripening, the seeds were rushed to Kew's conservation biotechnology laboratory to be planted. Unfortunately their worst fears were realised – the seeds had failed to germinate. But even though no new plants had been acquired, the production of the fruit was the evidence that the team needed to prove that the last café marron wasn't sterile. However, as only one in a thousand 'cut and stick' trials was successful, Carlos believed that some other factor must have caused this one plant to fruit when it did. On analysing every aspect of the conditions of the plant he recalled that the fertilisation had occurred amidst one of the hottest heatwaves that southern England had experienced in years, and at a time when the shades of the nursery roof had been broken, preventing them from closing. He deduced that the exposure to excess heat and light could have triggered the plant to fruit. Carlos began moving café marron plants into areas of exposed sunlight along the heating pipes that skirt the tropical nursery, and lo and behold many more plants began to fruit. As before, seeds from the fruit were rushed to the propagation units of Kew's conservation biotechnology laboratory, and a month later four out of five seeds had put out roots and shoots – the long-awaited lifeline for the next generation of café marron.

Carlos monitored the new plants as they began to grow, but quickly spotted that the plants that began to sprout did not have the lush, oval waxy leaves of the adult plant and its cuttings, but instead put out long thin brown leaves which looked almost dead. He was trying to propagate saplings from an elegant broad-leaved tropical tree and the plants he saw before him resembled something more like an ugly scrubby shrub. As he watched the saplings growing over a number of weeks, he saw their tattered brown leaves get longer and longer as the plants got taller, but then miraculously at around a metre in height their leaves began to change. Slowly their thin leaves plumped out, and astonishingly the adult plant revealed itself as the lush green tree easily recognisable as café marron. This amazing morphogenesis is thought to be a result of the plant's historical relationship with the native fauna of its island, the Rodrigues giant tortoise and a large bird called the solitaire, both of which are now extinct. As tortoises have very poor eyesight they rely on finding large green leaves to eat that are highly visible in their habitat. The thin brown leaves of the young café marron would therefore go unnoticed by a hungry tortoise, and it is only by the time they are well out of reach of the herbivore's straining neck that they produce their thick green, veritably appetising leaves. This strategy, called heterophylly, is seen also in the raintree (*Lonchocarpus capassa*) which grows in the savannahs and woodlands of southern Africa. Its leaves look grey and diseased to deter herbivorous antelope when it is at their head height as a small shrub, and only later, when it is a mature tree many metres high, does it produce lush green leaves.

Today the new café marron plants are producing their own flowers and their own seeds. From at first containing just five seeds, fruit are now being produced containing up to 85 seeds each, and slowly but surely a new population of genetically varied café marron is being re-established on Rodrigues, with the hope that once again it can flourish in its natural habitat. On the one hand this plant's adaptations to its environment remind us of the unyielding and inherent ability of plants to survive. On the other hand, the story of the café marron outlines the important work that continues to be carried out by botanical gardens and research institutes across the world, to help bolster populations and maintain the vital biodiversity of species on Earth.

The Power of Fungi

Chapter *Eight*

'For all the myriad forms that make up the diversity of plants on our planet, this amounts to just a small fraction of the life which exists in the world of fungi.'

Although they are not classified as plants themselves, the lives of fungi are so intrinsically linked to the success of plant life, that no understanding of botanical systems would be complete without acknowledging fungi for their crucial part in the evolution of life on our planet. But for the most part, the hidden power of fungi goes unnoticed. Those with little or no degree of biological inclination can appreciate the unfathomable diversity and seemingly endless beauty of the plant world, and even without an understanding of why plants exist in such beauty and variety, they are attractive to human senses of smell and vision and are therefore aesthetically pleasing. Fungi, on the other hand, make one think of edible mushrooms, odd yellow organisms such as lichens that grow on gravestones, perhaps mould, and a whole host of strange-looking toadstools. From early school days we learn that flowers are beautiful and important, while we're told that fungi are toxic and to be avoided, and for most of us this misunderstanding perpetuates into our adult years. But the real significance of fungi on our planet lies in what goes unseen.

For all the myriad forms of flowers, leaves, stalks, vines and fruits that make up the diversity of plants on our planet, this amounts to just a small fraction of the life which exists in the world of fungi. A kingdom unto itself, the world of fungi is made up of an enormous number of different groups, each defined by their unique shapes, sizes, behaviours and life strategies, allowing them to carry out a vast array of vital ecological roles. Unlike the cell walls of plants, which are made of cellulose, fungal cell walls are made of chitin, the same substance which makes the hard exoskeletons of insect bodies. And where plants store their energy as starch, fungi store their energy in the form of glycogen, the same energy storage molecule found in animal muscle cells. In their smallest forms fungi are single-celled aquatic organisms like chytrids, from the Greek *chytridion*, meaning 'little pot'. In their largest forms they are complex multicellular bodies which stretch for miles, underpinning entire ecosystems. In total there are believed to be between 700,000 and 5.1 million different species of fungi on Earth, and it is estimated that as many as 95 per cent of them are yet to be found, making them roughly six times as diverse as plants. But the reason that fungi are often mistakenly believed to be insignificant is largely due to the fact that they are highly inconspicuous in the environments in which they are encountered – despite their extreme abundance. In contrast to the flamboyant nature of flowering plants, fungi are less reliant on animals for reproduction, and consequently they do not exhibit extravagant colours or shapes on anywhere near the same scale. While some do produce large and colourful structures, many are shades of greys, light blues, creamy-whites and browns, comprising subtle structures tucked away out of view. Where plants have evolved large green photosynthetic structures to absorb as much sunlight as economically possible, fungi do not photosynthesise, and therefore their above-ground structures which we notice tend to be relatively small.

However, with fungi it is often what you don't see that is most surprising, and if you dig down into the substrate you will begin to reveal something of the subterranean lives of fungi. Beneath each mushroom lies a network of cotton-like threads called hyphae which act in a similar way to the roots of a plant to absorb

the nutrients which keep the organism alive. These structures are also important for the fungus's reproduction, territorial combat and the replacement of its competitors. When they amass together these threads can produce the fruiting bod-

above: Hidden power

Although out of sight, it is the mycelium which drives the amazing force of fungi.

—

ies above ground. These grow as toadstools and mushrooms, which are the spore-producing reproductive structures. Underground, however, the threads spread through the soil to create a mesh-like mat called mycelium. It is this mycelium that drives the vigorous advances of fungi. The tough and resistant nature of mycelium allows fungi to thrive in all habitats of the world, from the tropical forests of South America to the polar regions, binding the landscape with their networks of fibres. There are even extreme forms of microscopic deep-sea fungi whose finger-like growths extend through undersea hydrother-mal vents. As the mycelium grows through the plant matter that makes up the soil it releases enzymes which break down chemicals in the roots of plants as well as from dead plant material, and this is then absorbed by the fungi as food. The mycelial structures can hold up to 30,000 times their own weight in

Although just a small part of the entire organism, it is the spore-bearing
fruiting bodies, otherwise known as
mushrooms, by which fungi are most
easily recognised.

—

water and soil, and are the most efficient of all organs at breaking down organic material, making them indispensable to the creation and maintenance of the planet's soils. But unlike a plant's roots, which grow and branch into smaller and smaller structures, the mycelia grow and branch but then reattach to, and fuse with, older branches, resulting in the formation of a dense neuron-like web of growth. This network grows so densely that in just 16 cubic centimetres of soil there can be as many as 13 kilometres of mycelium, creating an extensive filter for the fungus's food absorption and gas exchange, like an externalised structure of intestinal villi or alveoli. The gaps which are created in this three-dimensional web of filaments create numerous tiny pockets in which water is stored, and these cavities also provide refuge to a whole community of microbial life. So not only does this network of threads trap nutrients, but it also gives structure to the mulch, helping it to resist erosion.

In plant roots and stems the structures of xylem and phloem allow them to pump nutrients hundreds of metres through the soil to the plant's organs, but fungi do not have xylem or phloem. Instead, many fungi have created their own transport system to match that of plants, which is perhaps best exemplified in the species *Armillaria solidipes* (formerly known as *Armillaria ostoyae*), a type of honey fungus. This common species from North America produces a 10-centimetre-wide typical umbrella-shaped mushroom, and is easily spotted growing in large tufts poking out of the ground throughout America's hardwood and conifer forests of the Pacific Northwest. But hidden from view, hair-thin filaments of its hypha-rich mycelial mat aggregate together, to form thick cylindrical cords up to 5 millimetres thick. These cords become enclosed in a hardened black outer coating, giving them the appearance of something like a tangle of black bootlaces. Called rhizomorphs, these bootlace-like structures act like a plant's roots, allowing the fungus not only to absorb nutrients but to carry them many hundreds of metres under the ground to wherever they are needed by the organism. With the growth of its black rhizomorphs, *Armillaria solidipes* is able to extend outwards in the soil over vast distances, wrapping around the roots of trees, growing under their bark and through rotting wood to absorb nutrients, and in this way it grows inexhaustibly under the forest floor, like an unseen behemoth. As a fascinating aside, the mycelial mats of some *Armillaria* fungi possess natural bioluminescence which makes them glow in the dark, radiating an ethereal blue-green light.

For a long time *Armillaria solidipes* was known to be a highly successful species, and in the forests where it grows it can cause a great deal of damage with its tough rhizomorphs. Foresters conducting research on the parasitic nature of this fungus would cut open dead pine trees and in most cases find underneath the bark a suffocating lattice of mycelium and black bootlaces tapping vital carbohydrates and water from the tree to feed the ferocious growth rate of the fungus. Research work carried out in 1998 set out to calculate the extent to which this fungus was responsible for the death of trees in an area of the Malheur Forest in eastern Oregon. Using aerial photographs of the forest, Catherine Parks, a scientist from the Pacific Northwest Station in Oregon, located areas where trees had been killed, and then set off on foot to take samples from 112 different sites of fungal activity. On comparing the DNA of the fungi from each site Parks was amazed to find that a total of 61 of them appeared to be identical. Even though these samples had been taken from parts of the fungus kilometres apart, they were actually all part of the same individual. By finding the outer limits of this one giant fungus Parks was eventually able to calculate its total size, and she found that it had extended over an astonishing area of 900 hectares, roughly the same size as 1260 football pitches. Occupying a depth of around a metre of soil, this colossal organism is thought to have a weight totalling at least 7000 tonnes, but possibly as much as 35,000 tonnes. Even at the smallest estimate, this fungus is without doubt the largest single organism on Earth. To have reached such a massive size it is estimated that it must have been growing continuously for as long as 2400 years, but perhaps as much as three times longer, meaning that the original fungal spores which gave rise to this organism may have been putting out their first hyphae at the same time as the construction of the great pyramids of Giza.

Although *Armillaria solidipes* is a fierce parasite of the trees of the Pacific Northwest forests, it still plays an important role for the ecosystem as a whole.

above: Spirit plant

The healing properties of *Ganoderma lucidum*, or ling zhi mushrooms, have been used in Chinese traditional medicine for thousands of years.

—

Although the success of the fungus is at the expense of individual trees, the removal of the older, more susceptible, trees creates gaps in the forest canopy, allowing smaller conifers and brushy hardwood species to grow there. This promotes a wider diversity of tree species in the forests, which is healthier for the habitat as a whole, and the fallen trees create food and shelter for a whole range of animal life – from microbial communities to small mammals. This process of renewal and recycling makes fungi vitally important.

Fungi are the planet's natural cement mixers. They drive the constant breakdown of rotting plant material, the bodies of dead animals and animal waste in the soil, which in turn releases nitrogen and carbon that can then be used by plants for growth. Usually these essential elements in the soil are locked up in large complex molecules, rendering them inaccessible to plants. Fungi help to break them down into smaller digestible molecules, which helps them to be absorbed through the cell walls of the plant's roots. Like the soil, the air is also full of an invisible mist of billions of fungal spores, and as soon as they land on the body of a dead animal or a dead plant, they set to work, growing and feeding. The key to fungi's ability to break down such a broad range of organic material is in the wide range of enzymatic secretions that they ooze from their bodies, including laccase and cellulase. These strong chemicals, called exoenzymes, allow certain species of fungus to break down large structures such as cellulose and lignin, which is the tough component of wood, and in turn transform them into sugars that the fungi can use for growth. If it weren't for fungi, forests would be overrun with millions of years' worth of dead wood, but instead fungi help turn this organic material into useful food for plants, while at the same time feeding the fungi themselves.

The diversity seen in the kingdom of fungi amounts to far more than just a useful addition to plant life. It is believed that the first organisms which ventured onto land from the ancient seas of the Precambrian era were fungi, in the form of lichens, around 1.3 billion years ago. These simple associations of single-celled organisms aggregated to form communities on the barren and rocky land, and as they grew their mycelium extended outwards into the substrate, slowly breaking it down. Oxalic acid and other acidic enzymes produced by the fungi reacted with the chemicals of the rocks, and in turn calcium and other minerals were liberated. As the enzymes combined with the calcium a chemical called calcium oxalate was formed, which helped dissolve the rocks and gradually break them into smaller pieces. Over a period of hundreds of millions of years this breakdown of the previously impenetrable surface of the land set in motion the first crucial steps towards the production of nutrient-containing soil. Without fungi taking these first steps to liberate minerals from the substrate, the first plants that came to colonise the land during the Palaeozoic era around 470 million years ago might never have become established. A recent study conducted by scientists at the University of Sheffield looked at the effect of fungi growing alongside extant relatives of some of the first land plants. It showed that the extensive 'greening of the land' was aided in part by a partnership formed between plants and fungi. For the study, thalloid liverworts, members of the oldest group of land plants, were grown in controlled conditions resembling the carbon-dioxide-rich atmosphere of the Palaeozoic era, and the results showed that these early conditions amplified the benefits gained by the formation of a symbiosis between the plant and the fungus –

plants which were colonised by fungi were able to absorb more photosynthetic carbon and benefitted by being able to absorb nutrients released by the fungi. As a consequence they grew much faster than liverworts that had not formed partnerships with fungi. In return the fungi received carbon produced by the plant. These results provide a fascinating insight into perhaps the very first partnerships which were forged between plants and fungi.

As the first land plants evolved to become vascular plants, and these diversified further still, fungi soon came to form associations with a wide range of plant species on whose roots they made their home. These types of fungi are called mycorrhizas. Primarily fungi use the plant as a source of food, mainly glucose and sucrose, which is produced in the leaves and then transported to its roots, and up to a certain level this is tolerated by the plant for the benefit that it receives in return. As the fungal thread-like hyphae bind themselves to the plant's roots they have the effect of vastly enlarging its surface area for water and nutrient absorption, and a mere 1-centimetre section of root can have its surface area extended to up to 3 square metres by the presence of hyphae. On top of this, the mycorrhizas selectively absorb particular minerals that the plant can use, such as phosphate, while excluding those that the plant does not require, such as sodium, and in this way they maximise the plant's uptake of useful ions. The mycorrhizas secrete chemicals which stimulate the roots to grow and branch, and to ensure their source of food stays healthy they secrete antibiotics which help protect the plant from harmful bacteria in the soil. Fungi are also able to capture harmful metals in the soil such as zinc, and so can allow plants to grow in soils that would otherwise be too contaminated.

Currently it is estimated that between 95 and 99 per cent of modern-day plants rely on a specific species of fungus for successful growth. But just as plants play a far greater role in our lives than the ecosystem services they provide, so too do fungi – in the form of medicines, chemicals, foods, drinks, as an aid in cleaning up pollution, and, more crucially, as fuel.

It is thought that as soon as early humans were developed enough to hunt and gather they would have collected mushrooms as part of their diet. Over thousands of years our ancestors discovered an increasing number of uses for the different fungi that could be found in the wild, and around 2500 years ago doctors and healers in China first began to document the incredible properties of various species. One of the most famous of the medicinal mushrooms detailed in the early Chinese medical texts is *Ganoderma lucidum*, known traditionally as the ling zhi mushroom or 'spirit plant'. This large fungus is found as a flat shelf-like growth protruding from the sides of trees on which it grows, and is easily recognised by its distinct shiny red lacquered surface. First documented during the Eastern Han Dynasty, it was used as a successful treatment for arthritis, asthma, high blood pressure, insomnia, and liver and heart disease, to name but a few, and its effectiveness while providing minimal side effects has earned it the reputation in Asia as the ultimate 'super herb'. The unique properties of this potent fungus continue to make it a highly valued medical commodity for traditional Chinese medicine, and it is most commonly used today as a treatment for degenerative and age-related illnesses, as well as having effective anti-cancer properties. China's long experience with the wild fungi of its forests has uncovered a great wealth of species with unique

above: Gills

Many fungi have gill structures under their cap which produce their reproductive spores in a shower of dust-like powder.

—

properties, which can be straightforwardly administered in teas and soups or simply eaten; a brew of the snow fungus (*Tremella fuciformis*)can heal ulcers and reduce fevers, a mixture of the wood-decomposing mushroom *Auricularia auricula-judae* with milk can be drunk to reduce throat inflammation and administered directly to soothe eye infections, and the straw mushroom (*Volvariella volvacea*), which is one of the mostly widely grown edible mushrooms in the world, can be eaten raw to help lower blood pressure.

Perhaps the most remarkable of the medicinal fungi that have historically been used in China is *Ophiocordyceps sinensis* (formerly *Cordyceps sinensis*) – known traditionally as 'winter worm, summer herb' – but more aptly referred to as the mind-control fungus. *O. sinensis* belongs to the largest division within the kingdom of fungi, known as the sac fungi, which includes some of the most important species as far as human use is concerned, including baker's yeast, antibiotics like penicillin, and the truffles. But what sets *O. sinensis* apart from these other forms is the gruesome way in which it reproduces. *O. sinensis* is what is called an entomopathogenic fungus, which means that in order for it to complete its life cycle its spores need to latch on to the body of an insect, which

above: Myco-restoration

Fungi such as *Pleurotus ostreatus* which can eat up toxic waste can potentially be used to regenerate polluted habitats.

—

will then carry them to a convenient place from which they can release more spores, and in doing so begin the process all over again. This is a trait common to many species of fungi, but *O. sinensis* differentiates itself in the way in which it exploits its arthropod host. In the grasslands of the Tibetan plateau large brown caterpillars of the *Thitarodes* moth feed on roots and grasses and other montane plants. The exact details of the moth's life cycle are still not fully understood, but it is known that when the *Ophiocordyceps* spores germinate they infect the young *Thitarodes* caterpillars, then become dormant inside their bodies for up to five years as the caterpillar grows. When the caterpillar is ready to pupate, the fungus starts to grow rapidly, feeding on the caterpillar's fat reserves, effectively starving it from the inside. Bit by bit the fungus mummifies its host from the inside out, and in its final death throes the caterpillar will climb up to the surface of the soil. Once the caterpillar is dead, the fungus fruiting body develops, bursting out of the head of the insect to form a 5-centimetre brown cylindrical mushroom that extends from the caterpillar's corpse to above the soil surface. The fruiting body then releases millions of tiny spores into the air, which float on the wind to find the next insect victim.

Although the parasitic *O. sinensis* terrorises the insect world, it is extremely valuable to humans as a medicinal fungus. Originally documented in Tibet during the late fifteenth century, the species has a long history of being collected for use in both Tibetan and Chinese traditional medicine; it has been highly prized for its ability to raise energy levels and aid recovery from illness, its hypoglycaemic effects, its ability to restore liver function and help reduce tumours. But the fungus remained a secret to the West until 1993, when it became well known following China's success on the stage of international athletics. At the World Outdoor Track and Field Championships in Germany, three previously unknown female athletes from China won all three medals in the 3000 metres, as well as gold and silver in the 10,000 metres, and just a few weeks later the same team broke three world records for long-distance races at the Chinese National Championships. Following these amazing performances their coach, Ma Junren, attributed their achievement to a regime of intense training and a diet of 'mineral-rich' soup containing *O. sinensis* mushrooms. Although Ma has since been the subject of doping controversies, the interest he generated in the powers of this miraculous species soon made it one of the most sought-after mushrooms in the world.

To ascertain to what degree this fungus can offer its supposed benefits, scientists have analysed the active chemical contained within *O. sinensis*, a

'Over thousands of years our ancestors discovered an increasing number of uses for the different fungi that could be found in the wild.'

previous: Bracket fungi

These wood-rotting fungi grow horizontally on trees and produce bracket-shaped fruit bodies that release basidiospores from tiny pores on their undersides.

—

compound called cordycepin, to determine its effectiveness as a treatment. Tests carried out on mice showed that cordycepin had an antidepressant effect, protected their digestive organs and bone marrow from harmful radiation, helped maintain blood-sugar and reduced liver damage – all from just one species of fungus. Extremely rare in the wild, *O. sinensis* was at one point only available to the Emperor in ancient China, and although its availability is no longer the reserve of monarchy and noblemen, it is still a highly valuable economic commodity in Asia and across the globe. Caterpillar corpses with their associated fungal growths which are collected from the wild are extremely valuable, and as global demand increases so does their price. Between the late 1990s and 2010 the market value of this fungus in Tibet has increased by more than 1000 per cent, and today a kilogram of large infected caterpillars with fruiting bodies could fetch over \$90,000 in the high-end retailers of Beijing.

The earliest part of mankind's study of fungi fell under the broader field of botanical study, owing to the fact that early authors believed them to be of the kingdom Plantae. But although they had not at this point been recognised as belonging to their own kingdom, fungi had long been claimed to have bizarre and unique characteristics, noticeably conveyed in the words of Lutheran botanist Hieronymus Bock in 1552: 'Fungi and truffles are neither herbs, nor roots, nor flowers, nor seeds, but merely the superfluous moisture or earth, of trees, or rotten wood, and of other rotting things.' The pioneering Italian botanist Pier Antonio Micheli was the first to undertake detailed study of fungal spores and their cycles of growth in 1729, but the Dutchman Christiaan Hendrik Persoon was probably the first mycologist. His 1801 publication *Synopsis Methodica Fungorum* contained the first Linnaean classification of a number of broad classes of fungi, thus creating the framework from which modern mycologists identify and classify fungi today. As with the plant world, the Victorian obsession with collecting, for reasons of both science and curiosity, drove the momentum for the study of fungal life forms in Britain in the nineteenth century, and in turn grand academic collections were founded to accommodate the growing body of work. To accompany their vast botanical archives, the Royal Botanic Gardens at

Kew founded their fungarium in 1879, to which were added the ever-expanding number of new species that were being discovered from across the globe. Today, over 130 years later, the collection is the largest and most important resource for mycological study anywhere on the planet, with a total of over

1.25 million individual specimens, drawing academics from all over the world to study the fungi and ancient texts held in the collection.

Today Kew's research into the lives of fungi is moving at a faster pace than ever, with the fungarium's unique resource allowing scientists to study the dried strain of Fleming's original penicillin culture, or species first discovered by Charles Darwin, alongside species discovered recently from the depths of the Amazon basin or the Asian steppe. This invaluable scientific resource will undoubtedly reveal yet more revolutionary findings about fungi in years to come. As it stands, fungi account for more than two thousand different edible species and almost five hundred with medicinal properties, and this relates just to the 5 per cent of fungi described so far. New species are discovered every year, and each new species has the potential to carry unique properties. It is prob-

above: *Agaricus
semiglobatus*

James Sowerby's *Coloured Figures
of English Fungi*, 1800.

—

ably no coincidence that more new species of fungi have been identified in the grounds of Kew than anywhere else in the world, as the gardens have a 200-year history of playing host to some of the world's foremost authorities on fungi. But this raises the question just how many new species of fungus must go unnoticed by the untrained eye. Every week new species come into the fungarium at Kew where they are described and catalogued, and a sample is sent to the lab for testing. There the fungus can be screened for the presence of metabolites that would have potential uses in medicine, cleaning up waste, in agriculture, or perhaps something else altogether. With continued momentum from centres of research such as Kew, the Centre for Agricultural and Biomedical Imaging (CABI) and similar scientific centres around the world, the potential for the discovery of powerful medicines and chemicals is high.

Fungi are hugely important to the healthy functioning of ecosystems – past, present and future – yet still remain poorly understood. This will hopefully continue to change as our scientific understanding of the plant world and the world of fungi continues to grow alongside one another.

One man who believes that such enlightenment is just around the corner is US mycologist Paul Stamets. Like many leading scientists in the field of mycology, Stamets has championed the incredible powers of fungi for many years, referring to something he calls 'myco-restoration' – the use of fungi to restore ecosystems. His vision is to exploit the natural abilities of fungi on a grand scale to absorb toxins and stabilise habitats, produce fuels and create super-medicines, and he already has 22 patents to his name for such fungus-related technologies. In 1998 he teamed up with a group of scientists from the Battelle research laboratories in Ohio to test the abilities of certain fungi to break down toxic waste such as from an oil spill. For their experiment they saturated four piles of soil with diesel fuel, and leaving one as a control they seeded the three other respective piles with enzymes, bacteria, and mycelium of a type of oyster mushroom called *Pleurotus ostreatus*, covering each of them with plastic sheeting. After eight weeks the plastic was peeled back, and the sight which greeted Stamets and the team was astonishing. While three of the piles were still blackened by the diesel and foul-scented, the pile seeded with mycelium had changed colour to a light yellow, and hundreds of pounds of oyster mushrooms had sprouted from the pile. The fungus had literally eaten up the diesel. The network of thread-like mycelium had produced natural enzymes called peroxidases which over the weeks disassembled the hydrocarbon fuel into its molecular components, carbon and hydrogen, which were then reassembled as carbohydrates and natural sugars, which the fungus could use to grow. With the covers removed from the piles they were left for a further eight weeks. As the fruiting bodies of the mushrooms came to the end of their life cycle they released spores which attracted hungry insects to the piles. While feeding on the fungus the insects laid eggs, and in time these hatched into larvae which also feasted on the fungi. Birds soon noticed these insect larvae and upon visiting the pile to feed they deposited seeds from plants in their droppings. Eventually these seeds germinated and sprouted shoots, and what was formerly a mound of contaminated fuel-covered soil was transformed by fungi into a green microcosm of life. Amazingly, when the pile was tested, the toxic component of the soil had been reduced from 10,000 parts per million to under 200 parts per million, in only a few months.

At least 120 novel enzymes similar to these peroxidases have been identified in other mushroom-producing fungi, and it is expected that these, in turn, will have important uses in breaking down pollutants and helping clean up toxic substances. Some of the most exciting species that are currently being examined for their world-changing properties are a handful of super-fungi which thrive on radioactive waste. These include numerous species which produce melanin (the pigment that gives our skin its colour) which have been found to harness radiation in order to grow. In 1999, a robot sent to Chernobyl returned with a sample of black mould that was found growing on the walls of one of the old reactors. Analysis of the growths revealed that in the same way that plants use chlorophyll to absorb light and make energy, these fungi were using their melanin to absorb ionising radiation to produce their own energy. This incredible ability, dubbed 'radio-synthesis', was a revelation. Prior to this discovery it was thought that only plants (and some bacteria) were able to produce their own energy independently of a food source. Even when starved of other nutrients, these fungi grew more than twice as large as other species when exposed to radiation, and it is thought that these types of fungi may have important roles in the future of food production. It has been suggested that translocation of fungal genes into plant crops could increase their productivity in harsh growing conditions.

Inspired by their successful trials of using fungi to clear up diesel fuel, the team from Battelle wanted to test the capabilities of mycoremediation on a grander scale, and so they set out to assess the ability of fungi to absorb the harmful waste that runs into watercourses from large-scale farming. They filled large fibrous sacks with wood and plant debris blown by the annual storms, and seeded the contents with fungal mycelium from three different species of fungus. They then sunk a layer of these sacks into the soil downstream from the farm, where animal waste runoff would have to pass through. When measured before it passed through the fungal filter, the levels of coliform bacteria in the runoff water (including *E. coli*) totalled around 172 colonies per 100 millilitres. But once it had passed through the sacks the powerful activity of the fungal antibiotics had slashed the number of bacteria to a relatively tiny five colonies per 100 millilitres – a massive reduction of 97 per cent in the level of contaminants. Other contaminants in the water, including nitrogen and phosphorus, were also noted to have been reduced after passing through the mycelium filter. Such a simple and inexpensive system could be used to reduce the pollution of watercourses and prevent eutrophication, and in the long run help maintain habitat stability in areas currently under threat from agricultural activities. Additionally, these setups use locally sourced materials and require relatively little maintenance to keep them running effectively.

The results from these experiments are exciting, and their implications are undoubtedly far-reaching, but for a mycologist they are not wholly surprising. They highlight the irrefutable role of fungi as environmental mediators: breaking down the raw materials of a habitat, and in turn building up a fertile environment for life to thrive in. Our scientific understanding of fungi is improving all the time, and we have only just scratched the surface of what secrets may be locked inside these incredible life forms. So far fungi have provided us with yeast to make bread and beer and wine, with powerful drugs, with food, and with tools to help protect and repair damaged habitats. As research continues and we discover more about these super-organisms, what we can expect to find will ultimately benefit all forms of life on Earth.

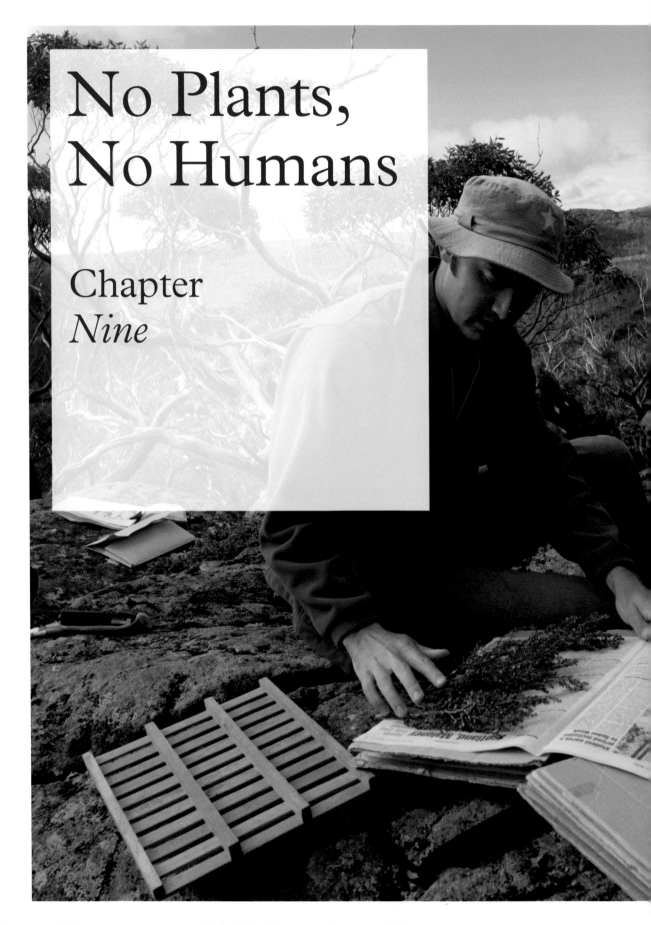

No Plants, No Humans

Chapter
Nine

'The increasing imbalance between humans and plants is poised to slip beyond the point of repair.'

P lants provide us with oxygen, and through this, we humans are intrinsically bound to the botanical life of our planet. Over millions of years they have given shape and form to the soil beneath our feet. Plants supply us with crops of fruit and vegetables and they give us a vast inventory of vital resources such as timber and cotton. Plants allow us the pleasures of beer, wine, caffeine and tobacco, and they sustain us with life-saving medicines and health-enhancing remedies. We are reliant on plants for almost every facet of our existence, and to this end we must be extremely careful not to lose them. But gradually we are losing species of both plants and animals from the web of biodiversity that makes up the life of our global ecosystems.

We know that plants are able to adapt to new conditions and venture into new and profitable partnerships with the world around them. But the changes to climate and habitat composition that face plants in the twenty-first century may be happening far too quickly for some plants to keep up, and as they fall by the wayside they are likely to slip off the world's species map for ever. As a number of plants begin to disappear from habitats ranging from the Brazilian rainforest to the Australian desert, African savannah and Asian steppe, we are set to potentially lose many of the important food and resource plants that people have come to rely on. And we are also losing the potential to realise the unique properties locked inside the DNA of those species that we have yet to discover. If we can live in equilibrium with the plant species of our planet, the global human population can expect to enjoy a long and prosperous future, one that affords it the many benefits that result from maintaining biodiverse habitats, such as natural resources, medicines and ecosystem services. But this can only be possible if we are able to strike a balance with the habitat requirements necessary for plants to sustain healthy and genetically diverse populations. The increasing imbalance between humans and plants is poised to slip beyond the point of repair. If it does, our human species may be heading towards a future where global prosperity, in the face of a quickening decline in plant species, becomes harder and harder to achieve.

Consider this scenario. If you went to see your GP with stomach pains or a headache that you couldn't shake, you would feel upset if the doctor spent the whole time analysing one of your feet, only to send you away with a full bill of health. A similar approach must be considered when we address what can be called the 'ecosystem illnesses' that are threatening the planet's biodiversity today. To study just one small cross-section of many of Earth's habitats may seem to show a healthy system in balance, of plants and animals flourishing in their habitat. But take a step back to see the bigger picture, and the truth reveals itself. A global census of the planet's plant life carried out in 2010, which aimed to assess the overall health of the planet's biodiversity, revealed that as many as one in every five plant species is currently at risk of extinction. While the pressures which threaten these plants come in the varying forms of habitat loss due to agriculture, pollution, over-exploitation and climate change, the unifying tone between them is that they are all the result of a human cause. Nearly 20 per cent of the planet's plant life could be lost in the coming decades if we don't minimise the effects of human activities on the natural world. Among that 20 per cent – totalling about 80,000 species – there may be a handful of obscure plants with little known value; they may not have specific relationships with any one animal in their habitat, they may offer no unique

ecosystem role and they may have no known economic value to humans. Their disappearance may go seemingly unnoticed, with their position in their habitat soon taken over by another species. But unknown to the world, one of those meagre plants

may contain unique chemicals that could reverse the growth of cancers, or even compounds with the ability to repair nerves damaged by Alzheimer's or Parkinson's disease. With less than 1 per cent of rainforest plants having been tested by scientists so far, these types of chemicals could very realistically exist. Unfortunately these losses can never be quantified.

However, the knock-on effect would be far more evident if one of the species that was lost were prominent in its environment, such as the mangrove trees of the Everglades. These plants provide a keystone community in their coastal habitat, where their extensive root systems not only provide shelter for a multitude of fish, crustaceans and bird life, but also act as a vital filter between the saline waters of the sea and the fresh waters of inland marshes. Furthermore, their presence along the coastline of Florida helps protect the land from wind and waves, and so prevents soil erosion. The loss of these tree species would mean the collapse of an entire coastal ecosystem, the communities of plants and animals that live there, and the degradation of around 800 square kilometres of habitat, with a high probability of a knock-on effect on habitats further inland. Urbanisation and agricultural expansion make an ecological disaster a very possible scenario, and we have already lost more than half of the world's mangrove forests in the last 50 years alone.

A vast body of knowledge has been gathered by botanists, ecologists and conservationists from botanical gardens and research centres around the world, which provides human beings with the vital tools to help us maintain the web of global biodiversity. This understanding of the interconnectivity of the planet's flora and fauna is crucial to ensuring that the most important species can be conserved. It also helps to ensure that the vital relationships between these key species can be maintained. Crucially, with an understanding of the vital ecological roles that each plant plays in its habitat, scientists are able to advise politicians and transnational bodies on how best to allocate global resources. Historical data can also provide an important insight into how we should or should not act in the twenty-first century if we wish to avoid the mistakes of our ancestors and in turn leave a legacy we can be proud of. For example, thanks to the plant collectors of the 1800s we know now that it may not be wise to simply bring plants from one country's habitat to another. Some plants can grow unchecked out of their natural surroundings and become invasive, and this is often not without consequence. Past examples include the irrepressible climber kudzu (*Pueraria lobata*) introduced to the USA from Asia, and the attractive white-flowered Japanese knotweed (*Fallopia japonica*) which has so far cost the UK over one billion pounds in eradication efforts. Thankfully, history also teaches us never to give up on any species, as we know that even plants that are reduced to just a handful of individuals, such as the Rwandan dwarf water-lily (*Nymphaea thermarum*) or the enigmatic café marron (*Ramosmania rodriguesii*), can today be bolstered with the hope of one day returning them to their habitats. But if we go back beyond the history of our textbooks, what is revealed is that the current impending ecological disaster is not the first time that the diversity of life on Earth has hung in the balance. Again and again, global catastrophe has shaken the very foundations of life, seeing plant and animal species go extinct in vast swathes

above: Fleshy fruits

To aid in their dispersal plants
package up their seeds in nutritious
sugary tissue of all colours, shapes
and sizes.

—

as they were unable to adapt to a sudden shift in conditions. But most importantly we must learn the lessons from what the fossil record tells us about how these drastic environmental changes ultimately affect life – that the species that go extinct are lost forever.

Since the evolution of complex life on Earth there have been five separate global events which have seen species disappear in vast numbers due to some form of catastrophic natural disaster. The first notable decline in species was around 440 million years ago, which saw as many as 25 per cent of animal families go extinct. Then during the Devonian period, some 370 million years ago, just under 20 per cent of animal families were lost, largely from the oceans. Around 251 million years ago the planet saw its most devastating extinction to date, with as many as 95 per cent of all plant and animal species on Earth wiped out under a cloud of noxious gas. Only 40 million years later, 23 per cent of families – including many mammal-like reptiles – perished, which ultimately made way for the dinosaurs. Since its discovery over 30 years ago, the fifth of these global catastrophes, which saw the dinosaurs wiped out by an asteroid impact around 65 million years ago, was believed to be the most recent mass extinction event. But as scientists have started to monitor the gradual decline of species in our modern era – most notably the beetles, amphibians, birds and large mammals – and analyse the results in the same way they would the ancient fossil record, what they are finding is extremely alarming. The Earth is once again heading into a mass extinction. Extrapolating upwards from the number of species which we have seen vanish in the last 500 years, scientists believe that we could be in the midst of the next great extinction event, which may continue for a further 300 to 2000 years. Given their habitat requirements, it is believed that the organisms which will most likely be worst affected will be the vertebrates. Habitat destruction and the disruption of their food chains will see their populations squeezed into smaller habitats with fewer resources, leaving them highly vulnerable with nowhere to escape to. This unfortunately has a knock-on effect on plants, as innumerable species have formed tight partnerships with these animals as their primary agents of seed dispersal.

The dispersal of a plant's seeds is a hugely important part of its life cycle. Through their evolutionary history this ability has allowed plants to colonise new areas and migrate in response to changing conditions of climate and com-petition. Seed dispersal is a highly important ecological process which affects both the dynamics and composition of plant and animal communities in all landscapes. The production of seeds was undoubtedly one of the greatest evo-lutionary innovations to have occurred during the radiation of land plants. The plant species we see around us today have evolved innumerable adaptations in their seeds, with each different form developed to suit a unique environment and life cycle.

The original seed producers, the gymnosperms, such as the extinct pteri-dosperms and the extant conifers and cycads, developed 'naked seeds' which do not have fruit tissue around them. As conifer cones dry, their scale-like bracts slowly open, releasing their seeds, which fall from the tree, floating on the breeze using wing-like structures to where they will germinate. Some of their relatives, including the ginkgos and yews, produce seeds with colourful fleshy coats, and these attract birds and mammals to disperse them. But it was the dawn of

above: Island of paradise

Cocos nucifera heavily laden with coconuts on the Madagascan island of Ile Sainte-Marie.

—

the angiosperms, the flowering plants, which saw some of the greatest modifications in the way that plants could spread their genetic material away from the parent plant. Fluffy seeds lighter than air evolved to carry seeds by wind power. Some, like the milkweeds, produced large pods from which numerous tiny seeds attached to wispy parachutes burst outwards into the air, while the feathery tails of clematis seeds allowed them to be carried on even the lightest zephyr. Others such as acers and sycamore trees evolved 'helicopter' seeds, which exhibit precision-engineered wings on which the plant's genetic material is carried, spinning away from their parent plant. In other species, the unique structures of their seeds can only be revealed at a microscopic level, such as those of the Indian paintbrush (*Castilleja flava*), which have a hollow honeycomb shape to create a large air-catching structure. Another strategy that proved highly successful was for plants to produce seeds squeezed tightly into pods, which, when ready, burst apart. A small mammal brushing past the ripe pod of the squirting-cucumber (*Ecballium elaterium*) will see it propelled up to 5 metres by the blast of exploding seeds, and the slightest knock to a pod of the Himalayan balsam (*Impatiens glandulifera*) will trigger a lightning-fast eruption of dark seeds in all directions. Some plants like poppies have salt-shaker heads, which throw out their seeds as they sway and bump into their neighbours in the wind, letting them fall to the ground. Or, in the case of many orchids, the minute dust-like seeds can be transported and carried off in a single puff of air.

But flowering plants did not limit themselves to using only the abiotic means of wind, water and jet-propulsion to spread their genes. As flowering plants have gradually radiated and become more diverse since the Cretaceous period, the role of animals has become increasingly important for their survival. The sticky seeds of *Galium aparine* evolved to adhere to birds and mammals that brushed past its gummy vegetation, and so too the hooked seeds of *Hackelia americana* became adapted to hitch a lift by grappling onto the fur of passing animals. These simple yet effective methods of seed dispersal proved very advantageous for the plant species which adapted to use them. Hand-in-hand with the evolution of flowers, a great number of animal species soon became necessary to transfer pollen and aid in their reproduction, and in time angiosperms also began to evolve specialised body parts to encourage animals to move their seeds, too.

All animals must consume nutrients to survive, and generally speaking those individuals which have better access to nutrients will be healthier and stronger and in turn will have a higher chance of reproducing. This is exploited by the plant world, and by offering their seed parcels to animals packaged in a nutritious coat, animals will ingest them and later deposit them in their droppings. Tasty, enticing and brightly coloured structures of sugary pulp evolved to encase the seeds of various angiosperms, drawing animals to feast on them and in turn transport them to a suitable place for their germination. This was the evolution of fruit. Over millions of years, natural selection drove the evolution of various types of these fleshy growths. The most appealing colours and tastes were best suited to attract foraging animals, and therefore provided the most effective seed dispersal service for the plant. In time these sugary parts became an important part of many animals' diets, giving rise to the planet's first frugivores.

By the time of the Tertiary, around 65 million years ago, many thousands

'Plants have evolved innumerable adaptations in their seeds, with each different form developed to suit a unique environment and life cycle.'

of different plants had evolved these rewards as a way of encouraging the dispersal of their seeds. This development is believed to have been tightly linked to the preferences of the dispersal agents. One hypothesis suggests that as the different species of frugivores became more numerous and capable, so too they promoted the evolution of larger, fleshier fruits – to maximise their foraging, fruit-eating animals would have selected the largest and most nutritious species, driving the selection of many fruits towards bigger, juicer and ultimately more nutritious forms. Simultaneously, other animals became more adept in foraging for smaller fruits, and through this the evolution of fruits radiated into a whole spectrum of specialised shapes and sizes. The result of this is evident today in the huge variety of fruits which can be found all over the world, with each one evolved as a unique vehicle to spread a plant's DNA through the specific composition of animals in its natural habitat.

opposite: King of fruits

The powerful odour of durian fruit makes it illegal to take on public transport in Malaysia, but it is still appealing to orang-utans.

—

Berries packed with sugar are most desirable for many different birds and small mammals. Some, such as damson (*Prunus insititia*), wild cherry (*P. avium*) and blackthorn (*P. spinosa*), contain just one seed each, whereas elder and holly berries contain many. Strawberries, raspberries and blackberries, misleadingly, are not berries at all, but are known as 'multiple fruits' made up of many single-seeded fruitlets. The huge hard-shelled durian fruit (*Durio* sp.) – the size of a bowling ball – is torn into by the dexterous hands of Asian primates, who discard the handful of nut-like seeds in its centre. The primates of the African forests make fast work of many hundreds of different fruit, including the likes of water-pear (*Syzygium guineense*) and the tennis-ball-sized mabungo (*Saba comorensis*), wild figs and the 50-centimetre-long fruits of the sausage tree (*Kigelia africana*), spreading the seeds of each as they travel through the forests and savannahs. Wild bananas (*Musa acuminata*) evolved to suit the taste buds of the short-nosed fruit bat (*Cynopterus sphinx*), and tannin-rich acorns have evolved to be distasteful to almost all wild animals aside from squirrels, which have the convenient habit of burying them in the ground. Pigments such as carotenoids, flavonoids and betalains give fruits their distinct colours, with the aim of luring keen-sighted animals, especially birds, and volatile chemicals give fruits their strong odours to

attract mammals who locate food using their noses.

But not all animals that eat fruit are economical seed dispersers as far as the plant is concerned. Some don't travel sufficient distances once they have ingested the seed and others may have strong chemicals in their digestive systems which kill the seeds before depositing them. There are also ravenous seed parasites such as snout beetles, who feast directly on the seeds of plants, destroying any chance of them germinating. And there are other animals that simply steal the sugary pulp from around the fruit, ignoring the seed altogether. To deter these unwelcome species many plants have evolved distasteful compounds in their seeds or in the flesh surrounding them. In fact there is even a theory which suggests that the first fruit structures to evolve may have been as a chemical defence for the seeds within. But while toxins make fruits and seeds unpalatable to inefficient dispersers, it also makes them toxic to potentially useful animal partners. It is, however, a commonly known behaviour for many birds and mammals to eat clay-like soils as part of their diet, a behaviour called geophagy, and the mineral structure of these soils helps to absorb plant toxins. Therefore the animals which include these soils in their diet are still able to eat and disperse even toxic fruits. Such behaviour is observed in the rhesus macaques (*Macaca mulatta*) of south, central and southeast Asia, who eat clay to absorb the tannins of various ingested seeds and fruits, as well as in various species of parrots which are found to lick the exposed soil along river banks as a way of neutralising the poisons taken up by eating too many fruits. In this way, even toxic seeds are still dispersed.

Today it is estimated that as many as 95 per cent of tree species in the tropics produce seeds that are dispersed by frugivorous animals, and every fruit, both extant and recorded in the fossil record, is the unique outcome of a plant's long evolutionary relationship with its animal partners. The fruits borne from these evolutionary partnerships exemplify the very fragility of this web, for there are many fruit species we see today which are no longer visited by their animal dispersers. These species are known as anachronistic fruit – uniquely evolved in shape and form, they exist today as mere echoes of a time when they were part of the diet of the giant animals of the Pleistocene that they evolved with, which have since gone extinct. The plains of North America were home to the mightiest animals since the disappearance of the dinosaurs 65 million years ago – such as mammoths, big cats and giant polar bears. South America was also home to a wealth of mighty megafauna, including enormous hippo-like toxodons, the giant ground sloth *Megatherium*, and huge tusk-bearing gomphotheres, all of which had thrived in these lands for tens of millions of years. Then around 14,000 years ago early human hunters arrived in these lands via a land bridge from Asia, wearing clothes made of sewn animal hide and equipped with stone-tipped spears. In a period of just a thousand years the megafauna of the Americas slowly disappeared. It is unclear whether this was due solely to the efficiency of these skilled hunters, but what is known is that by about 13,000 years ago these massive animals had been completely wiped out. Many plants began to decline, and in time they, too, followed the animals into extinction. Those species which relied on their fruits to be carried long distances from their parent plants would have been the first to succumb, while those that could survive with their fruits germinating close by, or could form new plants by putting out shoots from their roots, were able to sustain themselves in localised populations. As many of the fruiting plants of the Pleistocene had evolved in partnership with such huge animals, other animals were not able to fill in as their seed dispersers. As a result, where a handful of these fruiting plants still exist today they do so in extremely limited population ranges, and many of them remain perched on the brink of extinction, with humans as their only modern-day dispersers.

A telltale sign of these anachronistic species for the scientists who study them are trees that produce large fleshy fruits which fall to the ground and rot beneath the parent's canopy. In evolutionary terms it is a huge waste of resources to create large, pulpy fruits which are not eaten. If these plants' seeds were to germinate beneath the parent tree they would be in direct competition with it for water, light and nutrients. Of all of these anachronistic species perhaps the most recognisable is the avocado (*Persea americana*). This oily staple of salads and flavoursome dishes all over the world evolved alongside the giant sloths and mighty mammoths of the Pleistocene, where its golf-ball-sized seed was readily spread throughout South America. But in these habitats today only the seldom-seen jaguar will swallow and disperse the large seed of the avocado, and consequentially natural populations of wild avocado are now very rare. Luckily, in the case of this tree, one species has taken over as its primary seed disperser – *Homo sapiens*. Our love of this fruit has seen cultivated varieties spread throughout Asia, Europe and Africa.

Other fruiting species which hang on to their existence today in the absence of their seed dispersers include Kentucky coffee (*Gymnocladus dioicus*), whose large black pods require the grinding and scraping teeth of a large frugivore to help it germinate, and without them the species has been reduced to a few wild populations on the floodplains of the Midwest. Elsewhere in North America the same fate has befallen the long seed pods of the honey locust (*Gleditsia triacanthos*), the grapefruit-sized osage-orange (*Maclura pomifera*) and the sugary fruit of the pawpaw (*Asimina triloba*), while in Central and Southern America *Annona* and the sweet, fleshy papaya (*Carica papaya*) are equally affected. The scientists who study these anachronistic species have calculated the 13,000 years that these fruiting trees have existed without their animal partners to be roughly the equivalent of just 52 generations for these trees – a relative blink of an eye.

The Clovis people who first came to the Americas and hunted the Pleisto-

above: Ghosts of
evolution

With few of its original dispersers still
alive, the avocado relies largely on
human cultivation to maintain
its existence.
—

cene megafauna towards their eventual demise probably had
no idea of the consequences of their actions on the long-term
health of the American ecosystem. We know now that seed dispersal is crucial to maintaining the composition of habitats; for
example, research in the jungles of Panama revealed that with
the seeds of some species of tree, 99 per cent of those dropped
underneath their parent tree perished within a year, but those dispersed just 45
metres away had a better survival rate. We also understand now how dispersal
increases the exchange of genes between individuals, which in turn increases
the genetic diversity of the species as a whole. Having a wider gene pool better
protects plants against changes in their habitat, and on a timescale of hundreds or thousands of years this helps safeguard their wider populations in the
future. But even equipped with this knowledge, the situation in tropical forests
around the world today has an eerie echo of the Pleistocene extinctions. In the
Peruvian Amazon the illegal trade in bush-meat has seen a massive decline in
its populations of large seed-dispersing primates and other frugivorous mam-

above: Survival capsules

Seeds are the plant world's natural survival mechanism – if we can save the seeds we can save the species.

—

mals. Where the trees which produce the large seeds dispersed by these animals have begun to diminish in numbers, smaller wind-dispersed plants such as lianas have taken over. The same is seen in populations of tapir and agouti in Brazil, which both play an important role in the seed dispersal of endemic palms, but are both threatened by habitat loss and hunting. Similarly, in the thick forests of central Africa, where the seed-dispersing forest elephants play a vital role in maintaining the diversity of a vast array of large-fruiting tree species – some solely dependent on the elephants – the shrinking of their populations has seen a marked decline in the diversity of younger trees. What these studies ultimately tell us is that we must protect the populations of animals in these natural habitats if we wish to maintain their plant diversity. If we lose our primates, our giant tortoises and our elephants, we may also lose the plants they disperse, and in turn the larger habitats on which humans rely for resources.

While the crucial work of conservationists around the globe aims to intervene before any animal or plant species suffers the same fate as those of the Pleistocene, it is already too late for some. What is believed to be one of the rarest plants in the world is housed in the Temperate House in the grounds of the Royal Botanic Gardens at Kew. This remarkable survivor is called *Encephalartos woodii*, and it stands about 2 metres tall, with the distinct dark green feathered leaves and gnarled bark of a cycad. Only one male plant of this species – from which this plant was grown – has ever been discovered in the wild, located in 1895 growing on a steep, exposed outcrop in Ngoye Forest in Zululand. But by 1916 even that last remaining plant had been taken from the wild. With no female plant ever found, the male *E. woodii* plants in cultivation will never be able to reproduce naturally, and will never be able to produce seeds. Ultimately it is reliant on horticulturalists to sustain its existence via cloned cuttings. However, for the rest of the plant world, there is still more time, and the horticulturalists and botanists at Kew today believe that it is never too late.

While plants like the cycad *E. woodii* may have come to embody the vulnerability of many plant species around the world, there are other species held in the collections of Kew that embody the innate ability of plants to survive, even in compromised conditions. These species remind us that by their very nature, plants are expert survivors. These species should inspire us to persevere in our efforts to uphold the plant communities which make up our global ecosystems. Three such species are those of the pincushion plant (*Leucospermum conocarpodendron*), which is a large yellow-flowered plant from the Proteaceae family, a species of *Acacia* tree, and species of *Liparia* with sunburst blooms, all of which originate on the Cape of South Africa. In 1803, a Prussian ship called the *Henriette*, carrying a cargo of tea and silk, dropped its anchor just off the Cape of Good Hope on its way from China to Europe. While the ship gathered supplies, one of its passengers, a Dutch merchant by the name of Jan Teerlink, used his time to collect seeds from the array of flowers and plants which thrive on the Cape, subsequently detailing the collection of 32 separate species in his notebook. But on its return voyage to Europe, the *Henriette* was captured by a British man of war, and Teerlink's notebook, together with the seeds contained within, was taken into British custody. Making their way back to England with the navy, a number of documents from the *Henriette*, along with Teerlink's belongings, were eventually passed from the Admiralty to the Tower of London, where they were to remain for over 150 years. By 2006 the material from the *Henriette* had

above: Fruits on the
forest floor

Over millions of years animals have
developed intriguing relationships
with many different plants, resulting
in an interdependent web of flora
and fauna.

opposite: Super survivor

The miraculous tale of the South
African seeds discovered in Teerlink's
notebook, including the *Leucospermum conocarpodendron*, is
just one example of the amazing
resilience of plants.

made its way into the collections of the National Archives, where one day they were discovered by a Dutch researcher, who fell upon the curious red notebook and the 40 packets of seeds tucked between its pages. A few seeds from each of the 32 species were promptly sent to experts at the Millennium Seed Bank, and after an initial analysis by the team there, work was begun to attempt to enable the seeds to germinate. No one knew what exactly to expect, as the seeds had been kept for over 200 years, at varying temperatures and levels of humidity. As feared, the seeds of 29 of the species failed to respond to water or nutrients, but miraculously three of them began to send out shoots and roots – a handful of what were revealed to be *Liparia* seeds produced 16 healthy plants, while from what were found to be *Leucospermum conocarpodendron* seeds one out of eight germinated, and from a species of which there was just one intact seed a small *Acacia* sprang out of its sapling leaves. The exact identification of this last species remains unknown, and will eventually be determined when it reaches flowering age in the years to come.

That any of the seeds were able to germinate at all was nothing short of miraculous. Even seeds of the toughest known species of cereal would have been expected to die after having spent such a long time in these neglected conditions. The seed morphologists who worked on Teerlink's extraordinary specimens believe that the fact that many plants of the Cape have evolved to deal with regular fires in their habitat may mean that these seeds developed especially tough protective coats; this may go some way to explaining their durability. Whatever the cause, these seeds plant hope in the minds of conservationists.

The Royal Botanic Gardens, Kew has a rich and long history of plant collection, identification and research, and since the birth of modern botanical study Kew has been at the forefront of our human relationship with plants. As we enter into unknown times, where plant collecting is no longer purely an exercise in curiosity, but a vital conservation tool in safeguarding the survival of the human species, it is Kew that is once again leading the way.

Located some 50 miles south of London, in a rural setting of outstanding beauty, lies the epicentre of Kew's ultimate insurance policy against the loss of global biodiversity – the Millennium Seed Bank (MSB). The MSB is one of the largest seed banks in the world. It is the hub of a global partnership (the MSBP) that forms the largest ex-situ conservation project anywhere on the planet. The main aim of the MSBP is to maintain the diversity of Earth's plants for generations to come, by collecting and storing samples of their seeds. Should any species of plant ever go extinct in the wild, some of its seeds can be grown from the MSB repository, with the hope that in time they could be re-introduced into their natural habitat. Tucked away in Sussex, surrounded by a meadow of wildflowers, the great glass and metal exterior of the MSB building betrays few signs of the significance of the work which goes on deep in its interior. Long glass-walled corridors lead past laboratories of busy scientists, clad in white coats, analysing sprouting seeds and plant material under microscopes. Ecologists and seed collectors unpack new batches of seeds, ready for processing, and teams of international researchers from 50 nations use their specialist knowledge to identify and categorise the contents of jars of seeds from all over the world.

But while this hive of scientific activity is the product of one of the most future-gazing conservation projects ever seen, its inspiration stretches back over 100 years to the end of the nineteenth century. In 1898, two researchers called Horace Brown and Fergusson Escombe, who worked at Kew, began to study how seeds from different plants could be stored at cold temperatures, eventually publishing their work in a paper entitled 'A note on the influence of very low temperatures on the germinative power of seeds'. Their study paved the way for the thinking that a plant's natural 'survival capsules', their seeds, could in fact be stored viably in cold, dry conditions for long periods of time, and by the late 1960s a basic seed storage vault was built in the laboratories at Kew to do just that. This seed store was intended originally as a device for storing plant material exchanged between Kew and other centres of botanical study around the world, but as Kew's collection of seeds and plant material grew, a larger vault was soon required. By the 1970s they relocated their ever-growing collection of seeds to Wakehurst Place in

Sussex, built a custom-made cold-store for their collection, and set out on the first of a string of large-scale international seed-collecting expeditions.

As Kew's work in documenting and cataloguing the flora of habitats across the globe continued to grow from strength to strength, so too the work at the Wakehurst seed bank has picked up momentum; separate teams of seed-banking specialists and seed researchers were formed in 1981 to staff the growing facility, and in 1983 seed exchange programmes were established to document and preserve plants from the most threatened habitats in the tropics and arid areas of the planet. But the more that researchers and conservationists began to reveal about the state of global diversity and our human reliance on the plant world, the more obvious it became that a project of mammoth proportions was going to be required in order to maintain the health of the planet for future generations.

The number of mouths to feed on the planet is fast increasing – the UN estimates the global population to reach 10 billion by 2100 – and at the same time hundreds of plant species are already documented to have gone extinct. While the prospects of food shortages are becoming increasingly real, more and more people across the world have to rely on plants instead of animals for their dietary protein, further increasing our reliance on the plant world. Currently, as much as 80 per cent of the human diet is made up of just 12 different plant species: four tubers and eight species of cereals. Coupled with this, only about 7 or 8 per cent of plant species are regularly eaten by humans. This massively disproportionate reliance on so few plants, while disregarding so many species, puts humanity in a potentially very vulnerable position, as monocultures are known to be susceptible to disease and catastrophic changes in climate. It became clear to scientists that it was time to return the focus to the other little-known plant species around the world that could potentially be used to help support a growing human population. By the mid-nineties the work of the scientists and botanists at Kew's Millennium Seed Bank project had stepped into the limelight to offer humanity a major lifeline.

Conceived in 1995, the initial MSB project set itself what at first seemed, to many people, to be an impossible task – to collect and store in its seed vault samples from every single species of wild plant in the British Isles. Luckily, the long history of Kew's collectors meant that the group tasked with this target had a head start, with 60 per cent of the seeds already in storage. This, coupled with the fact that islands like the British Isles have relatively few habitat types, and so relatively few species, allowed them to exceed expectations, and in just a few years the seeds of a staggering 97 per cent of the 1800 or so species had been collected and stored in the MSB. The success of the project attracted wide media attention, not only to Kew's achievement but to the scientific relevance of such an accomplishment. In the years since, it has proved to be a crucial landmark in the struggle to preserve the planet's diversity. As the project gathered momentum, funding became available from a number of donors to begin the construction of a state-of-the-art storage vault at the Wakehurst site in 1998. This hi-tech facility for the study and storage of the world's seeds was specially designed to harbour the vital DNA material of plants from all over the world for many generations to come. Above the ground it appears like many other modern research facilities, but in its underbelly is its real *pièce de résistance*: down a glass elevator and across a large space-age atrium, a reinforced steel door leads through an airlock to the epicentre of the MSB project, its vast fortified seed vault. As its biological contents could one day hold the keys to the survival of the diversity of plant life on Earth, and ultimately the survival of the human race, it has been made to withstand every foreseeable calamity. The vault's metre-thick concrete walls can protect its contents from the radiation of a nuclear blast and are even tough enough to withstand a direct hit from an air crash. Once inside the first airlock, a dry room leads through several holding rooms into the heart of the seed vault: a number of giant cold stores set to minus 20°C, containing row upon row of thousands of glass bottles and jars, holding seeds from across the globe.

By 2000 the construction of the MSB's hi-tech seed vault was complete, and Kew's scientists entered the next phase of this revolutionary project – to collect seeds from 10 per cent of the entire world's flora by 2010 – an estimated 24,200 species. Almost as soon as building work was finished, the chilled shelves of the vault fast began to fill up with the MSB's rapidly expanding collection of seeds. As specimens began to flood into the Wakehurst site from 135 different countries, the total number of conserved species crept ever closer to that crucial 10 per cent mark. Finally, in mid-2009, with the help of more than 120 partnership organisations from over 50 different countries, the last of the 24,200 species was received. This landmark seed sample was from a pink variety of wild banana from China, which not only plays an important role in the diet of the Asian elephant, but is also an important genetic resource from which commercial banana crops can be enhanced. In turn, this seed sample too was cleaned, x-rayed, dried and banked in the MSBP vaults alongside the others, marking the completion of perhaps the most important step in preserving the world's flora to date.

For many of the seeds added to the collection it is unclear for how long they will remain viable, as no one has attempted to store them in this way before, and so periodically their viability must be assessed. To do this a sample of seeds is taken from the vaults and germinated on nutrient jelly, where their root and shoot growth can be monitored. If the seeds germinate successfully then their batch can remain in the vaults for a further 10 years, but if they fail to germinate sufficiently or are found to contain mutations, further samples must be grown and possibly new wild replacements will need to be collected. In 2012, Kew and its global partners are hoping to surpass expectations once again, and they have just recently set themselves a new target – to collect, safeguard and preserve samples from a further 15 per cent of the world's plant species by 2020. This incredible goal for the MSBP, to store seeds from a quarter of the world's plants, as a back-up for the individual collections held in their countries of origin, will no doubt result in one of the most valued collections of biological material anywhere on Earth. Every day the MSBP's collection is increasing, with around 3200 additional species filling the vault's empty shelves each year, including the addition of over 250 unidentified and possibly new species, unknown to science, in the last 10 years. These are perhaps the most significant figures in the project's work, as these are the plants for which we have sufficient genetic material to ensure that they never become extinct.

The ability to guarantee a future for as many plant species as possible is at the very core of the MSBP's objectives. Alongside Kew's endeavours, other major seed storage facilities around the world are helping to build the great 'library of plant life', including the mighty Svalbard Global Seed Vault, built 120 metres into the frozen mountainside on the Norwegian island of Spitsbergen. Unlike the MSBP, which focuses solely on wild species, the Svalbard vault, nicknamed the Doomsday Vault, includes the seeds of the world's most important crops, in case a global catastrophe should wipe any one out. Buried in ice and rock 1000 kilometres from the North Pole, even if its power fails this seed store will remain cold enough to preserve its collection, meaning that even in the unthinkable event of near-total apocalypse, these species will be saved. Collectively there are around 1400 seed banks around the world, and this incredible databank of genetic material could likely be the key aid to the survival of the human species in what currently looks like an uncertain future. In the cases of the five mass extinction events that the Earth has already seen, it has taken between 4 and 20 million years for biodiversity to bounce back to pre-extinction levels. Obviously we cannot wait that long to regain the species that will potentially become extinct in the coming decades, and these seed stores will be a major tool to prevent this as far as plant life is concerned.

above: Kew's Millennium Seed Bank

The bomb-proof seed vault could be our species' ultimate lifeline.

—

The incredible resource that these seed collections provide the scientific community, and the world at large, is undoubtedly one of the most forward-looking tools that we have to help maintain our planet's biodiversity. However, the foremost goal must be to preserve the health of every natural habitat on Earth as it is, before we are forced to dip into these lifelines of stored genetic material. After all, to re-establish any plant from stored seeds will require healthy soil, clean air, a stable climate in which they can grow, not to mention the associated plant and animal neighbours which constitute a healthy habitat to support them. As we know, these things are only available in healthy ecosystems. But the conservation of animals is even more complicated than that of plants. Unlike plants, we cannot simply take the eggs of the bees that pollinate our crops and put them on ice, and we cannot freeze the embryos of vital seed-dispersing elephants ready to defrost when current populations have gone extinct. We will not be able to re-introduce nectarivorous bats which spread pollen and seeds, and once they are lost we will not be able to reanimate the keystone primates which uphold their rainforest habitats. For these species there is no storage vault. Their only chance lies in the natural web of life in which they have evolved over millions of years – a web that both supports them and is supported by them at the same time.

So when we consider the diversity of life in the wild habitats of our planet,

above: Lychee

Cultivated for over 4000 years, the lychee also features in Chinese food therapy.

—

we must do so with John Muir's words on the fundamental connectedness of nature echoing in our ears: 'When we try to pick anything out by itself, we find it hitched to everything else.' The preservation of just one inconspicuous plant or animal may in some cases seem insignificant, but the relationships within the planet's webs of life are so complex that the loss of just one species can have a whole series of unpredictable consequences. The loss of a plant can mean the loss of an insect, the loss of an insect can mean that a bush loses its pollinator, and the loss of a bush can mean that a mammal has lost its food plant.

So as we continue our journey through what is believed by many to be one of the most critical centuries in the entire history of life on Earth so far, we must think of the legacy that our behaviour today will leave. The rate at which species are currently disappearing in the twenty-first century will send us headlong into the next global extinction event, perhaps within just a few hundred years. However, unlike every extinction event that has come before, this time there is no great fireball in the sky and no volcanic gases to blame, just the actions of one conscious, intelligent and incredibly powerful species. Human beings have a unique opportunity to avert the global catastrophe which sits distantly on the horizon. We have scientific understanding of an unfathomable magnitude, allowing us to cure diseases, create complex machines and explore deep into the cosmos, and we have a vast ecological understanding of the origins and inner workings of the natural world, from single-celled life to food webs composed of millions of organisms. We have all the know-how to measure the health of our planet, and we have all the necessary tools to fix it, thanks to the knowledge from botanical gardens, universities and indigenous groups. Every day the dedicated work of these groups of people provides an increasing body of findings to back the global preservation of species, and most importantly they are able to suggest effective and sustainable relationships between human society and nature. So gradually, the right steps can perhaps still be taken before it is too late. Ultimately, we find that the problems of habitat destruction, increasing greenhouse emissions and dwindling biodiversity are intrinsically connected to the driving forces of economics and politics, which in turn are linked to social issues of health and poverty. So as academics and grass-roots organisations alike are increasingly finding ways to make governments listen, we can expect to slowly reach the solutions which will benefit both humans and nature.

It was the exploits of pioneering explorers and plant hunters who first lit society's fascination with the botanical world in the eighteenth century, as they uncovered the unfathomable diversity of these organisms. And in the same way, in decades to come, we can expect that it will be the work of botanists and ecologists who will once again enlighten us and help us make the right decisions. There is no doubt that it will be a huge challenge for us to strike the right balance to secure the survival of the human species alongside the diversity of plants, and indeed all life on Earth. But luckily for us, we're one species that likes a challenge.

Index

Entries in *italics* indicate photographs.

blue morpho (*Morpho peleides*) 12, *12*
Bock, Hieronymus 214
Bolivia 175
Borneo 154, 162, 164, 189, 192
Boxall, William 39
Brazil 40, 149, 150, 179, 222, 236
Brown, Horace 241–2
bryophytes 22, *22*, 23, 27–8, 30
bryozoa 25
burdock (*Arctium* sp.) 149
buzz pollination 52, *52*, 53

C
Cactaceae 168
cacti 54, 62, 130, 150, 151, 154, 162, 168, 171,
     172, *173*, 174, 175, 178, 179, 181,
     182, 183, 184–5
café marron 192–4, *195*, 196–7, 224
Caladium steudneriifolium 137
calcium oxalate 207
California 61, 140
Californian lilac (*Ceanothus* spp.) 70–1
Californian poppies (*Eschscholzia californica*)
     185, *186*, 187
Cambrian era 18–19, 48–9, 60
camouflage 181, 185
Campbell, Archibald 37
Camponotus schmitzi (carpenter ant) 162
carbon dioxide cycle 14, 15–16, 17, 19, 81,
     84–5, 87, 89–90, 168, 172, 174, 207
Carboniferous period 27, 31, 48
carrion flowers 116
Catasetum 40, *40*, *41*, 44, *45*
Catasetum christyanum 40, *40*, *41*
Cattleya 39, 40, *40*
Cattleya skinneri 40, *40*, *41*
caudiciforms 172
cellulase 207
cellulose 25, 200, 207
Central America 40, 67, 185, 234
Chase, Mark 49
Chihuahuan Desert 174, 182
Chile 40, 179, 182, 184
China 36, 67, 208, 209, 211, 214, 237, 244
Chinese traditional medicine *105*, 107, *206*,
     209, 211
chitin 200
chlorophyll 15, 18, 64, 75–6, 83, 134
chloroplasts 18, 183
Chrysonotomyia ruforum 122

chytrids 200
cinchona tree *106*, 107
clematis 228
Cleopatra 36
climate 26, 32, 52, 59, 84–9, 167–8, 170, 172,
     185, 221–2, 227, 242, 245
Clovis people 235
cocklebur (*Xanthium*) 141
coco-de-mer 65
Cocos nucifera 228, *228*
co-evolutionary arms-race 164–5
Columbus, Christopher 99
columnar cacti 150, 181
communication, plant 110–27
conifers 25, 31, 52, 57, 66, 119, 203, 206,
     227–8
Cook, Captain 36, 143
Cooksonia 24–5
Copiapoa cinerea 181
Copiapoa laui 182
Corypha umbraculifera 65
crassulacean acid metabolism (CAM) 168, 172,
     174
Crematogaster (acrobat ants) 154
creosote bush 174
Cretaceous period 48, 50, 51, 52, 53, 73, 228
cross-pollination 56, 129, 143, 194, 196
Cryptostylis subulata 116
Cunningham, Alan 36
cyanobacteria 15, 17, 18, 19, *20–1*
cyanogenic glycosides 164
cyanohydrins 164
cycads 31–2, 36, 48, 52, 227–8, 237

D
da Vinci, Leonardo 76
Daintree Rainforest, Australia 60, 61
damson (*Prunus insititia*) 230
Darwin, Charles 39–49, 115, 136, 155, 215
Day, John 39
dead-horse arum 118–19, *118*
deception/trickery, plant 42, *42*, 53, 116, *117*,
     133, 135–7, 154–64, 181–2
Dendrobium 39
Dendrobium formosum 40, *41*
Dendrobium sinense 116
desert tortoise 154
deserts 32, 59, 60, 64, 130, 140, *148*, 150, 151,
     154, 166–87, 222
desiccation 22, 23, 30, 174, 178, 185

Acknowledgements

The insight into the amazing lives of our planet's plants described in these pages is the result of a great legacy of scientific endeavour from nations all over the world. As much as this book is a celebration of the diversity of the plant world, it is also a celebration of this global body of work.

Whilst it may not be possible to mention every scientist, researcher and specialist whose dedication has contributed to our current understanding of the botanical world, I hope here to thank all those key people who have directly provided their knowledge, understanding and expertise to help create this book.

Without the continued support of the team at the Royal Botanic Gardens, Kew, this book (and the accompanying television series) could not have been made. A great deal of the research, historical information and stories contained in these chapters came from my numerous meetings with the passionate horticulturists and botanists who work behind the scenes at Kew. Their generosity with their time and their help in bringing their work to a wider audience has been invaluable. I would especially like to thank Angela McFarland, Nigel Taylor, Carlos Magdalena, Monique Simmonds, Chris Ryan, Nick Johnson, James Beattie, Lara Jewitt, Wes Shaw, Scott Taylor, Marcelo Sellaro, Dave Cooke, Nigel Rothwell, Tony Hall, Steve Ketley, Katie Price, Richard Wilford, Kit Strange, Hannah Banks, Mark Chase and Phil Morris. Huge thanks must also be given to Kew's press and estate teams for tirelessly helping me locate the necessary people and plants during the research stages of this book. In this way, the continual help of Anna Quenby, Bronwyn Friedlander, Bryony Phillips, Tarryn Barrowman, Jo Maxwell, Dan McCarthy and Julie Bowers has been hugely instrumental.

As this book has been researched alongside, and written to accompany, Sky3D's television series *Kingdom of Plants 3D: With David Attenborough*, a huge thanks must be said to the incredible work of the fantastic team at Atlantic Productions who made this monumental project a reality. Special thanks must go to Series Producer Anthony Geffen, Director Martin Williams, Associate Producer Oliver Page and Editor Peter Miller. Naturally, the book has benefitted from those key members of the production team who helped make the whole series, particularly the contributions made by Tim Cragg, Rob Hollingworth, Tim Shepherd, Olwyn Silvester, Marie-Louise Frellesen, Charlotte Permutt, Anna Rayner, Ruth Sessions, James Prosser, Matt Baker-Jones, Aleksandra Czenczek, A. J. Butterworth, Jacquie Pepall, Skip Howard, Mimi Gilligan, Claudia Perkins, Andy Shelley and Ben McGuire. It has been an immense pleasure to work with such talented people.

Without Sky 3D's backing this series could not have been made, and for their continual support throughout, special thanks should go to Celia Taylor, John Cassy, Sarah Needham, Stuart Murphy and Sophie Turner-Laing. For the production stages of the book, thanks firstly to our book agent Jonathan Lloyd at Curtis Brown. At HarperCollins a large team has worked on this book, and I am especially grateful to Myles Archibald and Julia Koppitz for their part. Gina Fullerlove and Fiona Bradley from Kew's publishing team also played an invaluable role in overseeing the proofreading of the chapters and in supplying a great number of the photographs included in the book. Chapters were kindly proofread by a number of scientists at Kew, including Tim Entwisle, Mike Fay, Bryn Mason-Dentinger, Mark Nesbitt, Hugh Pritchard, Paula Rudall, Michiel van Slageren, Mary Smith, Brian Spooner, Martyn Ainsworth, Paul Cannon and Wolfgang Stuppy. A great many of the fantastic images provided were taken by Kew's photographer Andrew McRobb. Thank you also to Professor Stephen Hopper – Director of The Royal Botanic Gardens, Kew – for kindly agreeing to write the foreword to this book.

That I could even begin to tackle such a huge topic as the world of plants is thanks wholly to the ceaseless fascination of the natural world instilled in me by my parents, as well as ten very happy years spent working in the gardens of Baxters Farm.

Finally, and most importantly, my thanks and those of everyone involved in this project go to David Attenborough. His never-ending enthusiasm for the natural world and his commitment to communicating it to people of all generations across the world is what makes projects such as this a possibility. His lifelong work to cultivate an understanding of our planet's flora and fauna is a continual inspiration to all those who work with him.